KT-561-030

Also by Christopher Somerville

The Road to Roaringwater: A Walk Down
the West of Ireland

Coast: A Celebration of Britain's Coastal Heritage

The Golden Step: A Walk through the Heart of Crete

Somerville's 100 Best British Walks

Britain and Ireland's Best Wild Places:
500 Essential Journeys

Greenwood Dark: A Traveller's Poems

The Living Coast: An Aerial View of Britain's Shoreline

Somerville's Travels: Journeys through the
Heart and Soul of Britain

Never Eat Shredded Wheat: The Geography We've
Lost and How to Find It Again

Where to See Wildlife in Britain and Ireland

The January Man

A Year of Walking Britain

Christopher Somerville

BLACK SWAN

TRANSWORLD PUBLISHERS
61–63 Uxbridge Road, London W5 5SA
www.penguin.co.uk

Transworld is part of the Penguin Random House group of companies
whose addresses can be found at global.penguinrandomhouse.com

Penguin
Random House
UK

First published in Great Britain in 2017 by Doubleday
an imprint of Transworld Publishers
Black Swan edition published 2018

A CIP catalogue record for this book
is available from the British Library.

ISBN
9781784161248

Typeset in Palatino by Falcon Oast Graphic Art Ltd.
Printed and bound by Clays Ltd, Bungay, Suffolk.

Penguin Random House is committed to a sustainable
future for our business, our readers and our planet. This book
is made from Forest Stewardship Council® certified paper.

MIX
Paper from
responsible sources
FSC® C018179

1 3 5 7 9 10 8 6 4 2

Dedicated to my father, John Somerville –
and the man inside the man

Contents

June
Isle of Foula

July
St Cuthbert's Way

May (2)
Upper Teesdale

March
Nidderdale

April (2)
Lake District

September (3)
National Forest

September (2)
Sherwood Forest

April (1)
Lancashire Coast

February
Long Mynd

September (1)
Bardney Limewoods

January
The Leigh
and River Severn

August
The Wash

May (1)
May Hill

October
Norfolk Coast

Lyme Regis

December
Cley Hill

November
The Harroway

The January Man
(Dave Goulder)

*Oh the January man, he walks abroad in woollen coat and boots of
     leather;*
*The February man still wipes the snow from off his hair and blows his
     hands;*
*The man of March, he sees the Spring and wonders what the year will
     bring,*
And hopes for better weather.

*Through April rain the man goes down to watch the birds come in to
     share the Summer;*
*The man of May stands very still, watching the children dance away the
     day;*
In June the man inside the man is young and wants to lend a hand,
And grins at each newcomer.

And in July the man in cotton shirt, he sits and thinks on being idle;
*The August man in thousands takes the road to watch the sea and find
     the sun;*
September man is standing near to saddle up and lead the year,
And Autumn is his bridle.

*The man of new October takes the reins and early frost is on his
     shoulder;*
*The poor November man sees fire and wind and mist and rain and
     Winter air;*
December man looks through the snow, to let eleven brothers know
They're all a little older.

*And the January man comes round again in woollen coat and boots of
     leather,*
To take another turn and walk along the icy road he knows so well;
The January man is here for starting each and every year
Along the way for ever.

Author's Note

I loved 'The January Man' the very first time I heard it sung. It was around 1980, in the poky upstairs room of the Old Crown pub in Digbeth, in Birmingham city centre. Martin Carthy was the singer. I can see him now, his spare frame quivering, eyes closed, as he sang unaccompanied, slowly and with tremendous feeling:

> *Oh the January man, he walks abroad in woollen coat*
> *   and boots of leather;*
> *The February man still wipes the snow from off his*
> *   hair and blows his hands;*
> *The man of March, he sees the Spring and wonders*
> *   what the year will bring,*
> *And hopes for better weather . . .*

'The January Man' was so full of allegory, so elliptical and seasonal, that I took it to be a traditional song. I tried and failed to learn it. Then I forgot it. And it didn't recur to me until I began to plan a book of travels on foot in a great circle round the British Isles, each location chosen to suit a particular month. I hunted out Martin Carthy's recording of the song, and as soon as I heard his rich and reedy voice I knew that this mysterious month-by-month parable of 'The January Man' would be the cornerstone of the book.

Instead of being attributed to 'trad. arr. Carthy', as I'd expected, the song was accredited to one Dave Goulder. It had been composed in 1966 or thereabouts, and had been sung and recorded by just about everyone who was anyone in the folk-music world. When contacted, Mr Goulder was charm itself in giving me permission to hang my book on the rather magical peg of his song.

'The January Man' provided inspiration of another kind, too. Many people of my generation (I was born in 1949) have struggled to make the sort of emotional connection with their fathers that they have taken for granted with their own children. The profound and often long-hidden effects of the Second World War, the reticence of that generation of men, their stiff upper lips and sto-icism, were hard for their sons to break through. They offered a model of what it meant to be a man that set the bar unattainably high. For a start, they had saved the world from an evil fate. Follow that!

My father, John Somerville, was sent to the Royal Naval College at Dartmouth at the age of thirteen, and joined the Royal Navy when he was seventeen years old. He served in a destroyer in the Mediterranean through some of the worst and bloodiest months of the war, and never spoke about it. After the war he left the Navy to join Government Communications Headquarters (GCHQ), the secret establishment in Cheltenham, and never spoke a word about his work there. It was walking together that brought us closer. Walking promoted talking, and in the end produced a kind of understanding we came to value.

None of this is unique to my father and me. But the circular course of the journey I took, the cycle of the seasons, and the circle of a man's life so poetically expressed in Dave Goulder's song, all point to a completion that one never ceases looking for.

January

Oh the January man, he walks abroad
in woollen coat and boots of leather . . .

NEW YEAR AT THE Leigh, and Blacksmith Lane is six feet deep in floodwater. Roo and I can only stand and stare. King Severn has launched his annual invasion of the Gloucestershire village where we grew up. His summer bed lies under Wainlode Hill, more than a mile away across the fields, but now the houses along Blacksmith Lane have their toes in Severn's water. The flooded lanes are crazed with paper-thin ice. Bubbles stand frozen stiff in the ditches, and the roadside grasses are knotted together like lumpy dreadlocks under a milk-white skin.

Roo pauses at the edge of the flood. I've known him, my friend of longest standing, my brother in all but genes, since I was three years old. What does a gentleman of settled habit and sixty-something years do now? He forges in for old times' sake, crackling the ice as he goes. Beside him I wade, step for step, till the water is within a couple of inches of my wellington tops and the pressure is beginning to compress the rubber round my shins. We stand like two old herons, up to our knees in the floods. I bend and scoop up a parallelogram of fractured ice, thinner than a pane of glass. I grin at Roo, and touch it to my tongue.

Earthy, cold, metallic, drying the saliva like the touch of a sloe. A spoonful of floodwater slops over the rim of my right boot. Instantly it's 1957. I am seven years old, sloshing home along the road in gumboots full of flood-water, listening to them squelch. I stop by our gate at the top of the hill, hopping from one foot to the other, and take the gumboots off to reveal sodden socks, the grey wool dark and dripping. I tip my left boot up and pour out a stream of water onto the grass – a satisfying little cataract, brown and foot-warmed. It smells of cows and earth, but it looks like weakly made Ovaltine. What does it taste like? I take off the other boot. It's funny how the water comes right up to the top when my foot's inside, but disappears halfway down when I take it out. I reach a finger down into the boot and bring it up dripping. Into my mouth it goes. Mmm. Not very nice. Now I know something even Daddy doesn't know. I'm pretty sure my best and only friend Roo has never tasted floodwater either – although, knowing him, I wouldn't absolutely bet on it.

Our family came to live at The Leigh when I was three, in the flood winter of 1952. The Leigh, down in the flatlands beside the River Severn, was everything that the villages high in the nearby Cotswolds were not. It was remote, a modest community of small-scale farmers on a lane to nowhere. It was self-effacing, turning its back on the main road to Gloucester and Cheltenham. It was so unfashionable as not even to exist on the social map of Gloucestershire. No one went riding or hunting or waltzing at The Leigh. No one stamped about in gymkhana boots or admired themselves in shop windows. Scattered along the lanes was a handful of mixed farms – corn, cattle, geese – and one or two smallholdings. A church where some relation of Dick Whittington was said to have been baptized. A little primary school, destined only to last another ten years; a tiny shop in a crookback cottage. That was it for The Leigh.

We lived on a ridge at the entrance to the village in a draughty former vicarage, Hoefield House. Done out in dark brown paint and wartime blackout curtains, it was a gloomy lair when we arrived in midwinter. Mum soon had it cheered up with white walls and flowers and bright sofa covers. I had my elder sister Julia to squabble and snigger with and a big garden to play in, but only Roo for a friend. Roo lived the best part of a mile away in an ancient black-and-white farmhouse called Cyder Press, down where a straggle of red-brick houses and half-timbered cottages circled round a loop of lane. As soon as my mother judged me old enough to be out of doors on my own – probably at six or seven years old – I was pelting down to Cyder Press day after day, to escape with Roo into the promised land of the ridgy fields beyond. That was flood country, the January realm of King Severn at his maddest; a no-man's-land where you could chew sour cider apples from the abandoned orchards and taunt the cows into giving chase, where you could run unsupervised across the Big Meadow as far as your gumboots would let you. For a restless little boy with a super-heated imagination and a proper partner-in-crime to call on, it was pure heaven. And I could get to heaven on my tricycle in fifteen minutes, barring accidents.

*

When I got to school I found that all the other children knew what their fathers did. Fathers sold things in shops, or pulled out people's teeth, or ploughed the fields and scattered, or were away fighting for the Queen. My father was different. He drove off in the car every morning to work in Cheltenham. He came back home at night, mostly, unless it was one of the times when he was away somewhere abroad, wherever that was. What he did for a living, though, was a mystery.

'Daddy, where do you work?'
   'In Cheltenham.'
   'What do you do?'
   'I work at the office.'

But then almost everything to do with grown-ups was a mystery.

Roo and I decide to start today's perambulation at St Catherine's church, up on high ground, and to beat the bounds of our childhood along the village lanes and field edges as far as the floods will allow us. Wandering round the churchyard before we start, I get a shock. Almost all the folk who peopled my boyhood at The Leigh are here, neatly stretched out under headstones marked with information about themselves that I can scarcely credit. They seemed as old as the hills to me back then, but these dates say that they were in their twenties and thirties, most of them, when I first knew them – the age that my own children are now. The Quembys and the Theyers. The

Troughtons and the Poultons. The Westons, the Teakles, the Chandlers and the Freemans. I have hardly given a thought during my adult lifetime to these people who formed my view of what people should be.

As we set off from the churchyard towards the village, the January ghost of my father comes striding energetically towards us along the well-worn road in a heavy blue winter coat. Under the coat he's wearing one of his father's old suits with flapping lapels. His tie is neatly tied, his dark hair well disciplined, his black shoes impeccably shined. Some nonsense or other back at Hoefield House has made the family late for church, and Dad's sense of duty has driven him out ahead of his flock to walk the road alone, striding himself free of irritation. 'I remember how your father used to step out on his way to church,' says Roo, 'big long strides, all chin up and shoulders back. No one else walked like that.'

We pass the red-brick house where green-fingered Harry Wilks lived. Harry's wife had the beautiful name of Marguerite. Harry invariably made a clean sweep of the cups at the horticultural show; Marguerite provided the village with new-laid eggs, half a dozen a time in a brown paper bag. Just beyond the Wilks' house the tributary lane from the lower village comes in on the left. Opposite the junction is the big field where Roo and I once sat on the gate and watched a two-horse team harvesting with a reaper-binder. One of the horses was called Blackbird – Blackie for short – and wore a little straw hat with holes cut out for its ears to poke through. The reaper-binder had a clattering drum of whirling wooden bars that fed the harvest to the cutter and on up into the belly of the machine, which simultaneously ejected a line of neatly tied bundles of corn onto the ground. These were forked up into conical stooks once the field was cut, and left for a cart to gather before the rain set in. The Leigh must have been just about the last place in Gloucestershire to use such antiquated machinery and methods at that date. It was a different world down our lane.

From the junction by the harvest field my way home lay straight ahead, past the village school and up the slope. The crest of the ridge overlooked the main road, the A38, where Austin cars and AEC Mammoth Major lorries and Midland Red buses hurtled noisily along from Gloucester to Tewkesbury. On the ridge stood Hoefield House, its feet well out of the floods. I passed it earlier this morning, shut away behind large electronic gates in a cocoon of trees laden with mistletoe. We won't get a welcome there today. Instead, Roo and I turn our backs on the harvest field and trudge down the long road towards Cyder Press.

Riding my trike down the lane to Roo's, I'd usually meet the village's oldest inhabitant stumping his slow way along this stretch of road. Today I have learned from his gravestone that this lifelong farming man had the fine biblical name of Josiah Weston, but to us he was only and always Old Mr Weston. He had a red moon-face with white bristles, and invariably wore a floppy hat and a cowman's coat tied round the middle with binder twine. Roo reminds me now that he wore gaiters, too. Old Mr Weston had the most startling blue eyes and a gentle Gloucestershire voice as rich as cream. He lived his days not by the clock but by the light in the sky. Once he remarked to me, 'I go to bed with the sun, and I rise up with him.' I pondered that for a long time afterwards. My sister Julia had a season of wondering whether Old Mr Weston might not be God.

Roo and I stroll along, our wellingtons flapping companionably in rhythm. We pass tall red Prospect House where Miss Mauser lodged, a dignified elderly refugee spinster with a ramrod-straight back, all mousey brown in box-pleated skirts and old-fashioned jackets with square shoulders and tight sleeves. She rode an upright pushbike as tall and angular as herself, with a little puttering petrol engine fitted to the back wheel.

Beyond Prospect House the lane passes an orchard, beautifully replanted and maintained these days, overgrown with enormous unkempt cider-apple trees back in the 1950s. The Leigh had been a famous place for cider, and in locations around the village the farmers of yesteryear had planted more apple trees than corn or grass to supply the local makers. But those days were long gone when we came to live at The Leigh, their legacy the tangled orchards and an ancient cider press that sat on the front lawn at Roo's house, its immobile stone crusher as thick and clumsy as an ogre's cartwheel.

At the turn of the lane is the Fête Field, where once a year Mrs Paul and Mrs Poulton and Mrs Chandler bowled for a pig and guessed the weight of a super-solid fruit cake, while Mr Troughton and Mr Weston and Mr Theyer competed to back their tractors and trailers round a fiendish course of posts and ropes.

Opposite stands the old cottage, humpbacked, half-timbered, with its poky dark room that held the village shop and post office. The memories are flooding thick and fast now. Mrs Poulton gardening, up to her waist in a sea of green and purple cabbages. You couldn't charm Mrs Poulton with a smile. Her soft-voiced sister-in-law Miss Poulton shuffling behind the counter to unscrew a glass jar and weigh out a quarter of tongue-tingling clove sweets – 'Very com-*for*-ting, the winter mixture.' Bert Poulton, the postman, cycling on his red Post Office bike in capacious wellingtons with orange tops, his knees out at 90 degrees, fetching and delivering letters twice a day.

'Bert? No, he never catches cold. I make sure he drinks his sprout water, that's why.'

'What's sprout water, Mrs Poulton?'

'The water we boil our sprouts in, of course. What else did you think it could be?'

We turn the corner. 'Bert Poulton had a black bike for cycling to church on Sundays,' says Roo. 'Remember that?' I don't. 'Well, he did. Wasn't allowed to use Post Office property out of working hours. My older brother used to tell me he just had one bike and painted it black every Sunday and red again on Mondays. I believed him, too.'

Blacksmith Lane proves impassable. We turn and trudge back around the loop of the lower village, past Dan Theyer's farm where a horse stands with bent head in a farmyard that seems hardly to have changed in sixty years. Past Island Cottage, site of the old-fashioned dame school where my sister had the Three R's drummed into her by fair means or foul. Past Thin Mr Theyer's farm up on its rise of ground. Roo and I always had a healthy respect for Thin Mr Theyer, and it became a sight healthier after we decided to explore the Haunted House. A thrilling feature of The Leigh for its two juvenile adventurers was the existence of a few houses that lay in the fields outside the compass of the village road, well beyond the reach of tarmac or motor cars. At least two of these old dwellings were still inhabited by self-sufficient characters whom the twentieth century had scarcely reached. Another had decayed to a heap of damp bricks in a slough of mud and stinging nettles where we feared to go. The Haunted House, standing two fields away from the road on Thin Mr Theyer's land, occupied a status in between; it had been empty for years, but still retained its two storeys and a house-like shape under a half-unslated roof.

How we convinced ourselves the house was haunted I can't remember. Probably Roo's big brother Roger had spun us the mother of all yarns about it. Whatever the catalyst for the expedition, we found ourselves one afternoon under its red-brick wall, looking over our shoulders in a stew of guilt and bravado. Roo got through a downstairs window, and I followed him inside. The first room contained mildewed walls and great gaps in the floorboards. The second room contained a broken table. And the third room contained Thin Mr Theyer, with a cattle switch in his hand. What the farmer was doing in the old ruin I can't imagine – probably checking that no silly little boys were breaking their necks in his property. It must have given him an unpleasant jolt when we came peeping round the door. We had been terrifying ourselves with tales of the ghoulies and ghosties we might encounter in the Haunted House, so he certainly gave us a nasty shock. We legged it back past the table and the floorboard gaps, out of the window and across the field, Roo in the lead, till we reached the stile. Roo leaped over, and I hopped up onto the footboard at his heels. But Thin Mr Theyer evidently possessed the skill of silent running, because I suddenly heard his angry roar in my ear, followed by a crisp smack and a sharp sting on the bum.

I scrambled over and we ran like hell for Cyder Press and safety. The wicked fled. No man pursued. And I spent the next twenty-four hours anxiously wondering if I dared approach Bert Poulton to intercept the outraged letter I was convinced that Thin Mr Theyer was writing to Roo's father and mine about how he'd 'Caught Your Son Trespassing'. Needless to say, no such letter materialized, either by the morning or the afternoon post. When we met Thin Mr Theyer in the lane a day or two later he afforded us his usual nod and said nothing. Roo and I slunk by like two naughty pups, and gave the Haunted House and its guardian a wide berth after that.

Walking the lane today, we relive the encounter. We are still laughing about it when we come abreast of Cyder Press. The old house appears to lean with the curve of the lane as it always did, its black timbers seamed with age, its windows peeping out like sleepy eyes from their dormer hutches. Cyder Press was Roo's place, the hub of our wanderings, my home from home. You could creep right into the sitting-room fireplace, curl yourself into the inglenook seat with Sophie, the ancient cat, on your lap, and be safe. On summer nights you craned your neck and looked up the black crusty flue to where stars floated in the square frame of the chimney mouth. In winter you kicked off your boots in the kitchen and came to sit in the ingle and have your knee scabs softened and your chilly cheeks scorched by the fire. The logs spat sparks, and thick yellow ropes of smoke twisted up a chimney that had grown fierce and powerful. Cyder Press was the port of embarkation for adventure with my shipmate by flood and field, and a harbour to tie up in after storms at home or a shipwreck in the old coal canal at the end of the Big Meadow.

Today Roo stands grumbling in the lane, harrumphing at the changes his old home has undergone. The tree that contained our treehouse has been felled. We carved our secret symbols with penknives into its scaly bark, a simple and effective 'ZXZ' for Roo, a very impractical eagle crossed with a brace of lances for me. A memory surfaces – the pair of us hiding among the branches of the treehouse and shouting insults down at The Leigh's one and only village teenager as he cycled past in enormous gumboots on a bike with handlebars shaped like a buffalo's horns. In his fury at our cheek, he braked so hard that he skidded and went sideways into the ditch. I can see the poor youth, unable to get at us, shaking his fist and swearing vengeance as he remounted and wobbled away past Cyder Press. What happened then? He must have got us back – big boys invariably did. But how did he do it? Thump us? Give us a kick with his outsize wellies? I ask Roo, but he can't even remember the incident. It's so vivid in my mind's eye. But perhaps I'm wrong. Perhaps I've made it all up. Who, apart from the two of us, can now vouch for these scrapes and misdeeds of ours that took place out of our parents' ken? I link my arm in Roo's and feel even closer to my former brother-in-mischief, my only authenticator, as we come in sight of the village pond and the mouth of the space/time portal we knew as the Green Lane.

Here's a shock. Where has the Green Lane gone? They can't have filled it in, can they? 'Ah,' says Roo, 'you've never seen the flood bank, have you?'

The Green Lane was one of the communication trenches by which King Severn would make his stealthy advance towards the front line of the village lane. A rush forward when the pressure of flooding from Wainlode Hill had reached critical mass, and the lane and its houses would be captured. The house at Cyder Press had not itself been flooded in Roo's lifetime, but we didn't consider it a proper January if the water hadn't at least crept under his gate and halfway up the slope of the garden. As for the neighbours, they were lucky to get away without an inundation. Why The Leigh's villagers shifted themselves and their dwellings from the higher ground near St Catherine's to the notoriously flood-prone levels of the Severn is hard to fathom, but that's what happened some time in the fourteenth century. Perhaps the Black Death drove them away from what they saw as infected ground.

In any case, the local historians of The Leigh – in other words, all its long-term inhabitants – were adamant in the 1950s that it hadn't always flooded like this. After all, that Cyder Press has some Saxon walling, that herringbone bit next the lane, don't it? Those old Saxons wouldn't have built anything where the floods could get at it. When did all the flooding start? Well, that was back before Grandfather's time, when they built the lock gates at Gloucester, look. That conclusive *look* was our regional verbal tic, providing emphasis at the end of a sentence. Come out of that orchard, look. Farmer'll have your hide for that, look.

Some time in the last few years the householders along the lane must have raised an almighty fuss, because a ten-foot-high flood bank has been constructed all along the field margins. Roo and I climb to the top and admire the way it cradles the lowest edge of the village within its green protective arm. Beyond lies the gunmetal-grey ice sheet of the floods, broken by willows and hedge lines. The spreading Severn has reached the foot of the flood bank, and there it has been stopped in its tracks. But it still occupies the Green Lane, which lies directly below us. The old cart road curves away towards open country, ice in its skirts, a waxy sheen on its surface. We go down the bank and test the going, but it's hopeless. Water wells higher over the fractured ice at every stride, our welling-tons constrict, they'll drown if we take another step. The Green Lane is knee-deep at the first gate, deeper still round the bend. Our further walking plans have been washed away. No Big Meadow today for Roo and me. So we shove our hands in our pockets and kick the ice to pieces and stare our fill.

The Green Lane was no more than a quarter of a mile from end to end. It was a narrow little cart track, and very muddy. At some stage someone had tried to combat its muddiness by dumping several wagonloads of stone and broken bricks along the roadway. They made a wobbly surface to walk on, especially in thin-soled gumboots. The Green Lane was separated from the forgotten orchards and ridge-and-furrow fields by deep ditches on either side. Frogspawn floated on the murky water in spring, duckweed in summer. In those seasons the ditches exhaled a warm sour smell. But it was in the winter, when they froze over, that they came into their own. If the ice was pimpled and opaque, it was probably thick enough to slide on. If it was edged with pretty white whorls and gave out a soft groan at your first footfall, then it was thin enough to fall through. The dangerous thrill of taking a second step away from the bank, and the lurch of the heart – not all that unpleasant – as the ice sheet bent inwards and splintered and your boot went crashing through into icy water, have stayed sharp with me.

Is the old cattle trap still there at the far end of the Green Lane? Roo says he's pretty sure it is. We lean on the gate side by side and watch the twin ghosts of our boyhood selves go sloshing away down Memory Lane between hedges and the reflections of hedges, wading out of sight round the corner, wading back to 1957. The cattle trap was a hollow box of tarry timbers where a dozen bullocks could be corralled for drenching or dipping. The top rail of the trap was our moot hall. Roo and I would climb up and sit there leg to leg, gazing out across the Big Meadow and deciding what to do with the whole enormous day. Go sliding on the floods? Make a boat and sail to Tewkesbury? Go ghost-hunting in the Haunted House? Scramble through the bramble hedge into the derelict orchard to steal a hankie full of withered cider apples too sour for anything but pelting each other? Climb the willow pollards and try to cut our secret symbols in the iron-hard bark? A hundred choices for two giddy boys set free on their own in a fifty-acre field.

Flood time was boat-building time, after the ice had melted. One day we found a sheet of corrugated iron in the ditch alongside the cattle trap. 'Coo, look at that!' As we tugged it out and balanced it on the corner of the trap it clanged against the timbers, iron on wood, a manly sound. Roo ran home across the ridge-and-furrow to Cyder Press, and came back with four raspberry canes and a pocket full of green garden twine.

'Put them here, like this.'

'No, like this.'

'If we bend it up like this, and put those there . . .'

'You've done a granny knot.'

'No, I haven't.'

'Yes, you have. You need a reef knot, my dad said. Like this, look – right over left, left over right. That's what sailors do.'

'No, they don't.'

'Yes, they do. My dad says, and he was in the war. Look, it's really tight.'

'Go on, it'll float. Go on, get in. It will float, really.'

But it didn't. I climbed down onto the treacherous thing, and it turned straight over and snagged me underneath. My eyes were full of murk, there was mud and water in my mouth. Roo's boots swam into vision. The clumsy sheet pressed me down and then swirled away. My face heaved out into the air and I saw the skeleton twigs of a cider tree scratching across the sky. The iron boat had sunk. I threshed around and sat up in six inches of water and howled the floods out of my lungs. A drowned rat couldn't have been drownder, Old Mr Weston said when he met me scowling and slopping home.

It was the reef knots. Right over left, left over right was what I'd meant to do. Right over left, right over left – that's what I'd done. Dad wouldn't have made a mistake like that. Dad would have got it right. That's what dads did. Dads always squeezed toothpaste and Seccotine fish glue from the bottom of the tube. They went on long walks in the Brecon Beacons. They knew the right way and the wrong way. Straight-ahead dads with their parted hair and shiny toecaps, their navy-blue gabardines, their neat toolboxes and sorted desk drawers and secret war compartments in their heads that you mustn't rummage in. The pattern of men for little 1950s boys with their own gabardines and secrets and short-back-and-sides.

'What do you do at the office, Daddy?'

'I'm a civil servant.'

'But what's that?'
'It's what I do at the office.'

You couldn't see the office from the Green Lane or the Big Meadow, or from our house for that matter. But I knew what it looked like – a grey low-rise sea of Nissen huts behind chicken wire, sprawled awkwardly on the outskirts of Cheltenham's elegant Regency spa town. A few aerials stuck up out of the roofs. There was a man on the gate. When Dad's Morris 8 wouldn't start, we'd sometimes have to drop him off on our school run. He'd get out of Mum's car and stride away, shoulders back, chin up, towards the gatehouse and the pole barrier, looking determined but also strangely vulnerable, as though we'd caught him out at something. The Nissen-hut city swallowed him up. No signposts said 'GCHQ' back then.

The Big Meadow was a throwback, like The Leigh itself. It was a lammas meadow, traditionally farmed since who knew when? In winter King Severn made his royal advance and retreat, scattering largesse in his watery train – mineral-rich silt that settled in the drowned grasses. In spring the grasses grew like billy-o. So did ragged robin and marsh orchids, yellow rattle and meadowsweet and buttercups by the million. Snipe zigzagged away when you walked the marshy meadow. Lapwings flew overhead in black-and-white crowds, creaking in their peevish wild voices, flickering like old film as they suddenly tumbled earthwards. In summer the farmer cut the grass and let the brown and black cattle loose to graze the short stalks with gentle tearing noises. You couldn't get a cow to come to you like a dog, unless it was chasing you. But if you shut the cattle-trap gate and leaned on the moot rail, the herd would approach on the other side with infinite caution, keeping their bodies back and pushing their necks out towards you with soft exhalations of unexpectedly sweet breath. The boldest cow might even extend a pale rubbery tongue and rasp your outstretched palm, leaving a smear of saliva stickier than glue that you had to wash off in the ditch.

'Eeeurgh!'

Sometimes a cow would behave very rudely, poking its tongue right up each of its nostrils in turn, causing us to collapse with the giggles. Roo and I knew we weren't allowed to pick our noses, but we tried in vain to emulate the cows. Neither of us could reach our noses with our tongue tips, so we were even-steven there. There was one skill that my friend possessed, however, in which he held the whip hand over me, try as I might to copy him. Roo could clip his toenails with his teeth.

On the far side of the Big Meadow ran the disused coal canal from Coombe Hill, the boundary of our wanderings when we were seven or eight years old. Water voles dived in with modest plops as you approached. We chucked stones at them, and when we got older we shot at them with air rifles. They were rats, weren't they? Rats were Old Mr Weston's enemies, and Mr Troughton's, and Dan Theyer's. We were actually *helping* the farmers by knocking them off. So we rationalized our attempts at exterminating the Big Meadow's 'rats'.

The Big Meadow was a launching slip to freedom. We

must have run thousands of miles there. By the time we were nine or ten we had crossed its far boundaries and were exploring a wider world – down to the Red Lion under Wainlode Hill to fish in vain in the racing bend of the Severn, or over to Apperley under its tump of hill to buy Corona pop and Fruit Salad chews (eight for a penny) at Mrs Perry's dark little back-room shop. Something lasting was ignited in me: a pleasure in exploring, and a feeling of being at home in the outdoors.

Roo wasn't with me on the occasion I arrived at Mrs Perry's in a rainstorm and found she'd shut up her shop unexpectedly for the afternoon. Mrs Perry's copy of the *Gloucestershire Echo*, spotted with raindrops, lay on a stone by her front door. I decided I'd save it from the rain, just for her. What a good and helpful boy! If Roo had been there, perhaps he would have dissuaded me. Somehow I doubt it. What gave me the effrontery to climb onto Mrs Perry's water butt, scale her sloping outhouse roof and climb in through her upstairs window I can't now imagine. But I can remember the self-righteousness with which I told myself that I'd walked all the way to Apperley, I wanted my sweets and my bottle of limeade, and Mrs Perry wouldn't mind.

I crept down the stairs, laid the paper on a chair, lifted the sneck of the door into the shop part of the cottage and found myself in sweetie paradise. I unscrewed jars, opened boxes, weighed and measured into brown paper bags. Then I jotted down a list of everything I'd taken.

4 oz bullseyes
1 Mars bar
1 sherbet fountain
2 gobstoppers
4 oz sherbet lemons
1 pkt sweet cigarettes

It came to 2/4d. I piled up the coins neatly and wrote a note to go with them, something along the lines of: 'Dear Mrs Perry, Your paper was getting wet, so I brought it inside for you. I have taken the above sweets and left you the money.' And just before I left, I signed it, 'FROM AN UNKNOWN FRIEND'. I must have been reading Enid Blyton that week.

I let myself out, walked home soaking wet with cheeks a-bulge, and thought no more about it. But I didn't remain an unknown friend for long. A week later came the summons to Coombe Hill police station. Breaking and entering, boomed Sergeant Wheatley over my downcast head: a Very Serious Matter, young man. The sergeant, an imposing figure of authority, administered a ticking-off that had first my mother, and then myself, crying our eyes out. It wasn't till we got home that I discovered she'd feared I might be sent away to Borstal. I had to apologize to Mrs Perry in person, a sticky interview, and accept a parental ban on ever darkening her door again.

The twin ghosts on the moot rail are late for tea. They pull on sodden socks and squelching gumboots and descend from their perch above the flooded meadow. Hands in their shorts pockets, they turn their backs on the cattle trap and wade back down the Green Lane, back to their twenty-first-century personas of two old friends who lean together on a gate by the new flood bank.

I ask Roo if he can still chew his toes. It doesn't appear that he can. We make our way thoughtfully back across the fields to the church and our cars. A week from now, we know, the river will have withdrawn to its rightful bed in the bend by Wainlode Hill, leaving plastic bags in the willow branches and lank brown straws and twigs along the flood bank. A scummy tidemark along the verges of Blacksmith Lane will be left to mark the limit of its advance this year, until the new leaves and fresh grass of spring have hidden and expunged all signs that King Severn has ever been there.

February

*The February man still wipes the snow
from off his hair and blows his hands . . .*

IT IS TURNING OUT a proper winter. I can't believe my good luck. The floods of January are over and done, and now the crack of February brings snow, a great dump of the stuff all along the Welsh Borders. I'm headed north on the outward stage of this long year's circle round Britain, and though the business of getting about is going to be a thundering nuisance for the next day or so, the fact of the snow will make up for that.

There is no walking to compare with walking in snow. It is transcendental cleansing, walking in order to walk away from oneself, the rhythmic *creak-creak* of boots on snow drawing the mind away across the blank white canvas of the countryside and up through the blank white air. This is the zenith of solo walking. I sit down and plan a route that will be up in the air in every sense, one that I've had in mind for years, a snowy walk across the roof of the great Shropshire whaleback named the Long Mynd. And although I'll be walking it on my own, I'll have a constant companion at my shoulder: the Reverend E. Donald Carr, Rector of Woolstaston in the north-east skirts of the hill, whose twenty-four-hour fight for life through a ferocious Long Mynd blizzard in 1865 is a heroic tale remembered far and wide along the Borders.

After rummaging in vain on shelves and in boxes for my copy of *A Night in the Snow*, the little book Mr Carr wrote about his ordeal, I have managed to get hold of

THE JANUARY MAN

another copy. As I read Carr's remarkably modest account of what befell him, I watch the Shropshire weather for signs as keenly as a jealous lover. It's going to be clear skies, promises the BBC; maybe some snow flurries, hazards the Met Office. Maybe some hill fog. Cold to start with, then the temperature rising. Rapidly thawing. They don't really know. I look out my walking poles, my thermals and my thickest gloves, and I remember the matchless thrill of waking up as a boy at The Leigh in the sure and certain knowledge that it had snowed in the night and was still hard at it.

The road that Hoefield House overlooked, the A38, was the main route that linked the West Country with the Midlands. Before the M5 motorway was opened in 1971, from spring to autumn the A38 was a conveyor belt of vehicles slowly grinding from Plymouth and Bristol up to Birmingham and Derby. Their roar was heard faintly from our garden, and caused my parents to sleep with the bedroom windows closed. The lane that trickled round The Leigh saw almost no traffic – it was a cul-de-sac leading to a closed circle. But at Hoefield House we were close enough to the A38 to be served by mobile vendors. In those pre-supermarket days the baker, the milkman and the fishmonger brought their wares to us in their small old vans with spoked wheels and thin tyres. Onion Johnny in his blue beret would lean his onion-draped bike on the gatepost and peel a couple of brown globes from the onion string he wore round his neck – a great piece of theatre – when selling to my mother. From time to time a knife grinder set up his hand-operated wheel in the verge and scraped away rather hopelessly at our kitchen blades. Tramps would push their ratty old prams full of kettles and cast-off bedding up from the main road to beg a spoonful of tea. There were a lot of tramps on the roads in those days, men with tangled hair and bad coughs, burly-looking in their multiple coats padded out with newspapers. Many of these shadow men must have been casualties in one way or another of 'the war', that entity that hung malevolently in the ether, inescapable to grown-ups, impenetrable to children. The tramps were mild-mannered men and presented no threat in my eyes. But the paraffin man was a different beast – a folk devil to my overheated imagination, although I never spoke a word to him. He had a triangular float hung with forks, corkscrews, potato peelers and fire irons that clanked with a sound suggestive of gibbets. A serving can with a fat hollow spout dangled from a brass hook at the back, and a sloshing sound and stink of kerosene came forth from the bowels of the float when its owner moved off. Thirty years later, watching the film of *Charlie and the Chocolate Factory* with my own children, the sinister Tinker and his jangling cart of knives and cleavers brought my dread of The Leigh's paraffin man back to mind with disturbing clarity.

Eleven months of the year we could expect the world to come off the A38 and pay us a call. But when the snow fell and the main road was blocked and the village lane rose level with the hedgetops, then silence arrived. The outside world creaked and plopped, our leaky gutters grew five-foot icicles that hung from the eaves like a giant's shaggy eyebrows, and Jack Frost crept into my room in the middle of the night to etch amazingly detailed shapes of ferns and acanthus fronds on the window over my bed. In the morning I would wake to subaqueous light in the room and a white flickering at the edges of the frozen pane. A penny warmed in the armpit for a minute or two and applied to the glass would melt a peephole and confirm the brilliant news.

'It's snowing! Daddy, it's snowing!'

The room where I slept had a dual function. For twenty minutes, early every morning, it doubled as Dad's dressing room. It was a mysterious concept, the dressing room. It was like having a potato-peeling room, or a dog-patting room. The function didn't seem important enough to have a whole room reserved for it. Nevertheless, Dad had one. Men of his background did. Dad's dressing room contained a curvy-fronted wardrobe, faintly smelling of dust and mothballs, where he hung his suits – and his father's suits, the 1920s ones with spivvy lapels that he occasionally wore because he couldn't bear to waste them by throwing them away. Under the window furthest from my bed stood a solid old chest of drawers where Dad would dress himself briskly for work, his back turned, his pale bare legs protruding from under his shirt tail, cursing the snow under his breath.

To me the slanting curtain of snow outside heralded a day off school. It spelt snowballs and sledging and sliding down the whitened road with Roo. To Dad it meant trudging down the slope to see if the snowplough had cut a passage along the A38. If it had, he'd have to face a tedious and dangerous drive to Cheltenham in his small 'going-to-work' car, a second-hand Morris 8 in austerity black, registration number GYC 898. So he'd bend and grunt in the snow outside, fixing a clumsy cat's cradle of snow-chain round either front tyre and barking his knuckles. 'God . . . blast it!' The cold engine would sulk as he pulled the starter, producing a self-pitying coughing that wheezed and faded like a dying consumptive. 'Oh . . . bloody hellfire!' A frantic scrabble in the boot for the starting handle. More fumbling at the front of the car as the fiddly little lugs refused to engage in their slot. 'Get *in* there, you . . . !' Several furious swings of the handle and more scraped skin ('Ow! Damn it!'), a spluttering roar from the engine, and with a reddening hankie tied round his knuckles he'd chuck the handle onto the passenger seat and slither the Morris down the drive.

'What job do you do, Daddy?'
    'I'm a civil servant.'
    'But what's that?'
    'Someone who works in an office.'
    'But what do you do all day in the office, Daddy?'
    'I work hard. Now, no more questions, eh?'

He couldn't consult anyone. The phone lines were down, so he couldn't phone the office. The office couldn't phone him. He couldn't ask his wife's opinion – she didn't know anything about it. Bad things were happening abroad.

People were relying on him. He could have stayed at home, but he wouldn't let himself. He had to get through. It was his duty, simple as that.

I rattle through the snowy Welsh Borders on the stopping train to Church Stretton, looking out at the white-capped Shropshire hills and thinking of Dad and the Revd Donald Carr. When the worst snowstorm in living memory struck the Long Mynd on 29 January 1865, it caught Mr Carr out in the open. He, too, had a conscience that told him he had to get through. Unlike Dad, though, he had no snow-ploughs or snowchains to help him climb the mountain. Donald Carr struggled alone and on foot through the deepest snowdrifts so that the tiny hillside community of Ratlinghope could have their Sunday service, and his sense of duty would have cost him his life on the way back to Woolstaston if he hadn't also possessed a will of iron and a faith that could move mountains.

To his regular duties at Woolstaston, Mr Carr had added the responsibility of seeing to the spiritual needs of the Ratlinghope folk, purely because no other clergy-man could be persuaded to accept such a poor, remote and miserable living. Week in, week out, in sun, rain, fog or snow, he crossed the Long Mynd on horseback or on foot to take Sunday service at the little church some four miles away on the far side of the hill. He was proud of never having missed a single week in nearly nine years of ministry at Ratlinghope. Even though conditions were atrocious that midwinter Sunday in 1865, he felt obliged to make the journey. 'I was anxious to get to my little church if possible,' he wrote later, 'in fact, I considered it my duty to make the attempt, though I felt very doubtful whether I should succeed.'

As soon as he had finished morning service at Woolstaston, therefore, the rector swallowed a quick mouthful of soup and set out for Ratlinghope. The snow-drifts soon proved too deep for his horse, so he sent it home with his servant and continued on foot and alone. It took him more than two hours to cross the hill, at times crawling through the drifts on hands and knees, and when he got to Ratlinghope he found that no one had really expected him. He conducted a hasty service for a handful of worshippers in the church. Afterwards several people offered Carr a bed for the night, but he refused. 'I was anxious to get back to Wolstaston in time for my six o'clock evening service.' How he thought he was going to manage that is hard to imagine. But duty's call was strong, and it was duty that saw him setting out again up the hill at about four o'clock in the afternoon, an hour before nightfall, without stopping for a bite to eat or a rest.

Within an hour the conscientious clergyman was in deep trouble. 'A furious gale had come on from E.S.E., which, as soon as I got on the open moorland, I found was driving clouds of snow and icy sleet before it . . . The force of the wind was extraordinary. I have been in many furi-ous gales, but never in anything to compare to that, as it took me off my legs, and blew me flat down upon the ground over and over again.'

In spite of the buffeting, Carr gained the top of the

Long Mynd as night fell. He still felt confident of getting home, thinking that he knew the hill like the back of his hand. But the frequent tumbles, the blinding snow and his own tiredness had combined to disorientate him. Instead of steering north-east into the lane to Woolstaston, he headed too far to the east, only discovering his mistake when the ground began to slope away in front of him. 'Suddenly my feet flew from under me, and I found myself shooting at a fearful pace down the side of one of the steep ravines which I had imagined lay far away to my right.' He tried to check his slide by using his walking stick as an ice-axe, but something jerked it out of his hand and spun him round. He fell headfirst down the slope, 'expecting every moment to be dashed over the rocks at the bottom of the ravine'. By hooking his leg into the snow he succeeded in stopping his slide; he turned himself the right way up, and got down into the bottom of the cleft, only to find himself 'scratching and struggling' in twenty-foot snowdrifts. 'It was now dark. I did not know into which of the ravines I had fallen. The only way by which I had thought to escape was hopelessly blocked up, and I had to face the awful fact that I was lost among the hills, should have to spend the night there, and that, humanly speaking, it was almost impossible that I could survive it.'

Donald Carr had fallen into Long Batch, one of the loneliest and steepest clefts in the flanks of the Long Mynd, over six hundred feet deep and extremely steep to climb out of even on a fine summer's day. It looked a certainty that he would die there in the snow during that freezing night. But the Rector of Woolstaston was made of sterner stuff than he himself knew.

Jolting out of Church Stretton station yard in a taxi, I crane forward and survey as much of the Long Mynd as I can see. The slopes rise steeply, their upper regions white with snow. It hasn't melted yet. But time is ticking on. When we get into the narrow hill road to Picklescott the taxi starts to slide around the icy corners, but we manage to arrive in one piece. The fire in the bar of the Bottle & Glass Inn looks tempting, but it's already afternoon. I want to cover as much ground as possible before the light begins to fade, and I have a nasty suspicion that the day has already run away with more time than I can really afford. On with the cold-weather coat, the hat and gloves and scarf, the boots and gaiters, and I waddle off up the lane like the Michelin Man. The fields here, almost a thousand feet above sea level, are milky green with melting snow, but the hilltop covering still looks solid enough. Somewhere up there under the cold blue sky I'll meet the Portway, and that ancient track should guide me south along the heights of the Long Mynd to where the Revd Carr conducted his stubborn fight for life.

For years I've been aware of the broad old trackway that pursues a zigzag course along the spine of the Long Mynd. From time to time I've used sections of it to knit circular walks together. But it's only recently that I've noticed the legend 'Portway' blazoned alongside it on the OS Explorer map in the elaborate Fraktur calligraphy that betokens an antiquity. The Portway, the 'way along which things are transported', turns out to have been in use for some five thousand years, judging by the stone axe-heads that have been found along its course. In medieval times it was a through route for drovers taking beasts north from Bishop's Castle to Shrewsbury market. From the seventeenth century onwards, ox-drawn wagon teams were able to climb its (marginally) improved roadway, and cattle drovers brought their herds this way from South Wales. Some knew it as 'the King's Highway on Longmunede', others as Via Regalis, the Royal Road. When the nineteenth century brought metalled roads and railways to the surrounding valleys, the upland track declined and fell into disuse. You can trace its course on the map with a fingertip, as I did, all the way down the length of the Long Mynd, a broad old road for travellers in rough times who wanted to keep their goods and animals high, dry and out of trouble's way.

A sweet foggy steam of cattle breath exudes from a corrugated tin shed beside the lane, where the beasts are tugging mouthfuls of hay from a rack. They are up to their hocks in mud, but they have that look of slow contentment that feeding cattle wear, swinging their heads very deliberately up and down as though nodding to the passer-by. The weather is hinged between winter and spring. A flurry of snow whips across the hedges, each fleck dissolving the moment it touches the ribs of meltwater that come rippling down the lane. I stop in a gateway to admire the sunlit hills dolloped with snow, the Long Mynd uplands billowing like a new sail, Caer Caradoc and the swaybacked Lawley across the Stretton Valley, a lone white shark fin a dozen miles away in the north-east that must be the Wrekin.

If I'm steering in the right direction, that black line of hedge at the ridge ought to mark the course of the Portway. Proper snow lies underfoot as I climb, a wet layer that becomes progressively crunchier. It's ankle deep by the time I reach the hedge. Can the margin of this loose run of twisted hollies and stunted thorn bushes really be the Via Regalis, the old royal road over the hills? If there's a hollow in the ground, beaten out by centuries of boots and hooves, it's well hidden under the snow. But the bearings, north-east to south-west, seem right. It's all rather disorientating up here, with little intense snow showers cutting across and everything in the landscape reduced to black and white. I'm normally a prolific generator of heat when walking, and wear as little as I can get away with. Now I'm hot and sweaty after hurrying uphill in all these winter layers. My glasses keep fogging up as warm breath condenses on cold lenses. When I push my hood back to cool my hot face, the snow wriggles in between scarf and neck.

An image from Donald Carr's travails in *A Night in the Snow* comes suddenly to mind. After the benighted rector had managed by superhuman exertion to climb out of Long Batch ravine, he promptly fell all the way to the bottom of the next one. He bashed and bruised himself all over. He lost his hat and his stick, and worst of all, his outer gloves of warm fur:

My hands . . . were so numbed with cold as to be nearly useless. I had the greatest difficulty in holding the flask, or in eating snow for refreshment, and could hardly get my hands to my mouth for the masses of ice which had formed upon my whiskers, and which were gradually developed into a long crystal beard, hanging halfway to my waist. Icicles likewise had formed about my eyes and eyebrows, which I frequently had to break off, and my hair had frozen into a solid block of ice. Large balls of ice also formed upon my cuffs, and underneath my knees, which encumbered me very much in walking, and I had continually to break them off . . . It may seem absurd to mention it, but the cravings of hunger grew so keen, stimulated as they were by the cold and the great exertion, that it actually occurred to me whether I could eat one of my old dogskin gloves.

I compare this with my own small discomforts, and burst out laughing. When do you really know you're in a tight spot? When you can contemplate eating your gloves. Dogskins especially.

The wind is rising out of the west, whistling shrilly in the holly trees that stiffen the backbone of the hedge. Ancient trackways are often bordered by hollies, the poor relations of the tree family, universally regarded these days as a nuisance and a waste of space. The drovers valued them as deterrents to straying, the beasts being reluctant to push their sensitive noses and vulnerable eyes through such a prickly screen. One or two of the Portway's hollies are so chunky and twisted I wonder whether they have grown distorted through pollarding. Some of the farmers on the Long Mynd in former years would painstakingly cultivate a hollins or holly grove to provide their cattle with sustenance in the winter months when snow covered the grass. Holly leaves are rich in nutrients and high in calories, and if cut from the upper section of a tree they tend to be less prickly. The majority of animals browse at low level, and it is these more vulnerable leaves that the holly invests with most prickles. It's rare to find an intact hollins these days; modern manufactured feed supplements are simpler to deploy, less time-consuming and more nutritious, though far more expensive. The old groves are almost invariably neglected, the unpruned trees grown so tall and tangly that it would be a tricky and painful business to harvest their uppermost leaves. There is one splendid specimen of a hollins, though, rather prosaically known simply as The Hollies, away to the west in the high ground towards Wales, which was bought by the Shropshire Wildlife Trust in 2008 after an appeal. When I walked there on a cold day near Christmas shortly after the acquisition, I found a rowdy gang of mistle thrushes robbing the winter landscape of its colour as they stripped the scarlet berries from the trees.

The Portway climbs gradually beside the long dark block of a conifer wood. A buzzard wheels above it, silent and intent, looking for prey in a world turned white and unforgiving. By the time the old road has reached the exposed summit of Wilderley Hill it is masked in shin-high snowdrifts. The snow has a glue-like tenacity; it coats my gaiters and ices my bootlaces. The old road burrows under the even white blanket of an open field, and I plod across it with heavy steps that crunch and squeak, making for a horizon indistinguishable from the sky. In his masterful account of his Land's End to John o'Groats walk, *Journey Through Britain*, John Hillaby speaks of the 'skull cinema' that operates in the heads of all walkers after a certain number of miles, the stream of fantasy, conjecture and connections that floods the mind while the body tags along behind, step by unconscious step. This afternoon, with direction-finding more or less guaranteed by the south-west trend of the Portway, I am free to sit back in the skull cinema and watch grainy black-and-white images of the Revd Carr as he stumbled half frozen in the dark across the hills, rolling and tumbling into one snowbound cleft after another, stretched to the limits of exhaustion and panic, grimly determined not to succumb to the temptation to lie down in the nearest snowdrift and fall asleep.

At one point during his ordeal the rector noticed a small shadow flitting about on the snow. As he approached, it diminished to a single dark spot. 'I put my hand down upon this dark object to ascertain what it could possibly be, and found that I had got hold of a hare's head!' The little creature had burrowed in the snow for warmth. There are hare tracks ahead of me on Wilderley Hill this afternoon. Long ovals have been dinted by the leg bones, with a pair of smaller circles made by the forepaws just in front, showing where the hare had hopped steadily along. Now comes a patch of kicked-up snow, followed by a resumption of the tracks, more widely spaced and with the circular indentations now slightly behind the ovals as the hare took fright at something and dashed away, its powerful back legs shooting forward of its front ones at each stride.

I don't catch even a glimpse of the fugitive hare, but there are plenty of sheep about. They huddle in the lee of the Portway's hedge, a flock of tattered beldames in long ragged fleeces that flutter in the wind. Four of them leave the warmth of the group and set out across the field in single file, moving uneasily, their black faces nodding up and down as they hobble pathetically along. It's a Bruegel scene in black and white. The sole colour contrast is between the virgin white of the snow and the fleeces of the ewes, stained a urinous yellow through a long winter's exposure to the elements. When the breakaways reach the far side of the white square, they find themselves open to the worst of the weather on the windward side of the hedge. They move jerkily on, bobbing and limping, unable to settle, still in thrall to whatever inscrutable sheepy imperative made them quit their sheltered situation and their flock fellows in the first place.

The wind plucks up snow devils and brings them whirling by, obscuring the sheep. Between one flurry and the next, the air suddenly clears towards the west and I catch a glimpse of a widely spaced row of outlines peaking along the ridge, the naked black extrusions of the Stiperstones. These jagged tors are of sandstone, baked and squeezed into quartzite 500,000,000 years ago, a spread of time easy to write, impossible to conjure with.

The contrast between their rudimentary, blasted-looking shapes and the smooth line of the ridge they rise from has given vent to a swirling cloud of demonic tales and beliefs about the Stiperstones. Edric the Saxon was an actual historical figure, a Mercian thane who burned Shrewsbury in 1069 as an act of defiance against Britain's new Norman rulers. But in Stiperstones mythology he is Wild Edric, a ghostly warrior betrayed by his fairy wife, who visits the Stones every winter solstice for a wild hunt with the witches and warlocks of Shropshire. Godgyfu, 'gift of God', later spoken of as Lady Godiva, was the wife of a Mercian earl, Leofric, and a tireless benefactor of poor and oppressed people. Her celebrated naked ride through Coventry, undertaken to win a wager with her husband and force him to relax his oppressive tax regime, is not necessarily a complete fiction. But the Godiva who is condemned to ride a spectral horse for eternity along the Stones by night is something else – a soul in torment, shrieking and wailing as she bemoans her wickedness in going hunting on the Sabbath when she ought to have been in church. The Devil gets the lion's share of the Stiperstones stories, of course – how the Father of Evil created the Stones when he let fall an apronful of rocks, how he almost snatched Slashrags the tailor off to Hell when they met by the Stones (Slashrags spotted Old Nick's cloven hooves and forked tail just in time), and how on misty days he sits on the largest of the outcrops, the Devil's Chair, waiting for England's ruin as the Stones sink back into the ground.

Shropshire novelist Mary Webb mined the rich seam

of ominous Stiperstones legend to great effect early in the twentieth century. 'For miles around [the outcrop] was feared,' she wrote in *The Golden Arrow*. 'It drew the thunder, people said. Storms broke round it suddenly out of a clear sky. No one cared to cross the range near it after dark . . . Whenever rain or driving sleet made a grey shechinah there, people said, "There's harm brewing. He's in his chair." They simply felt it, as sheep feel the coming of snow.'

Lead seams in the ridge were worked from Roman times for the best part of the following two thousand years. It's hard to be sure how many of the highly coloured legends were invented by the lead miners. Like the hill farmers, those rough, uneducated men knew the Stones and the bleak countryside they dominate, year in and year out, in the harshest weather and blackest conditions. As incumbent of Ratlinghope, only a couple of miles from the ridge, Donald Carr knew the lead miners of the Stiperstones. It is his story I have been projecting onto the screen of my imagination as I follow the snowbound Portway along the Long Mynd; and now a fresh curtain of snow sweeps across the hill, shutting away the black pinnacles of the outcrop and bringing the programme in the skull cinema to an abrupt end.

It's cold. It's really windy up here, and getting windier. The first hint of nightfall is creeping into the sky. My legs are beginning to ache from the heavy walking through snow. I glance at my watch. My God, it's four o'clock. I take a look at the map. Where am I? Somewhere up above Betchcott – four or five miles from Picklescott, if I'm going to complete the circuit I've set myself. That's not going to happen before dark, is it? The snow has slowed me up, far more than I'd bargained for. Come on, be realistic. Turn round and go back to where you crossed the road. You can drop down to Picklescott in under an hour from there.

I experience a curious reluctance to leave the hill. To leave the shadowy figure in the snowdrifts that my imagination has been fleshing out all along the walk. What would Donald Carr think of me? Oh, come on. You can resume this walk tomorrow from this very spot. Click on the skull-cinema projector and he'll be right there. In the meantime – how about the Bottle & Glass, and that lovely fire of theirs?

In the morning I push back my bedroom curtains at the Bottle & Glass to find that grey cotton wool has filled the world. I can't even see to the far side of the car park. The denseness of the fog and the drip of trees and gutters tell me that the forecasters' thaw has well and truly got under way. I put my head outside, and find that the wind has dropped. It's ten degrees warmer than yesterday – shirt-sleeves weather. It's also unwalkable weather, at least for anyone who wants to see further than the end of their nose. I open the map and retreat into *A Night in the Snow* to see if I can work out exactly what happened to Donald Carr during the latter half of his odyssey in the blizzard of 1865. Tomorrow, as far as I'm concerned, can look after itself.

After a day and night of gentle wind and solid hill fog,

of rushing streams and trickling lanes, the next day dawns a little clearer. I can see the treetops and the vague grey loom of the hills. My suspicions are confirmed once I've climbed the long lane from Picklescott to the crest of Wilderley Hill. The fog has lifted up here, to reveal a sodden green landscape from which the snow has all but melted away. Drifts of wet snow still linger along the verges of the Portway, but the track itself is now awash with puddles. The sheep make little sucking sounds with their hooves as they trot off across the muddy fields where the grass lies flattened into untidy hanks by the now vanished snow. It's dank and damp, the mizzly air pearling my coat with miniature drops. But as I follow the old road across the Betchcott Hills the sun makes a watery appearance ahead, and the temperature begins to dip again.

At a place the map names Duckley Nap a road crosses the line of the Portway. As far as I can reckon, it was here or hereabouts that Mr Carr took his wrong turning. A few hundred yards of squelching to the east across the heather and peat, and I find the ground beginning to fall away. I'm looking down into the shadows of Long Batch. A few patches of snow still cling to the hillside. It is a fearsome slope. Even in the most favourable conditions there'd be a fair chance of losing your footing while descending. On a glassy crust of wind-smoothed snow a tumble would be inevitable, and there'd be nothing to prevent you shooting all the way to the bottom.

I retrace my steps to the Portway, and return to the black-and-white feature film in the skull cinema. The other batches or ravines in the vicinity of the Long Batch are equally steep. How on earth did our hero get into and out of so many, one after another? How did he fail to break his neck, or his legs, or bash his brains out on the rocks that floor the batches? Somehow he survived the long night of 29 January, falling down, scrambling up, hopelessly lost. 'Never did shipwrecked mariner watch for the morning more anxiously than did I through that weary, endless night, for I knew that a glimpse of the distance in any one direction would enable me to steer my course homewards. Day dawned at last, but hope and patience were to be yet further tried, for a dense fog clung to the face of the hill, obscuring everything but the objects close at hand. Furthermore, I discovered that I was rapidly becoming snow blind.'

Unable to feel anything with hands or feet, incapable of distinguishing more than shadowy shapes or seeing further than a few yards in any direction, Carr wandered here and there through the fog in a desperate attempt to keep awake and maintain his circulation. His hat, stick and fur gloves were gone, and now he somehow lost his boots as well.

They do not seem to have become unlaced, as the laces were firmly knotted, but had burst in the middle, and the whole front of the boot had been stretched out of shape from the strain put upon it whilst laboriously dragging my feet out of deep drifts for so many hours together, which I can only describe as acting upon the boots like a steam-power boot-jack. And so for hours I walked on in

my stockings without inconvenience. Even when I
trod upon gorse bushes, I did not feel it, as my feet
had become as insensible as my hands.

The Portway leads on south-west from Duckley Nap, and
brings me to a corner of road from which a post in the
heather points the way into Light Spout Hollow. The path,
streaked with the last vestiges of the snow, becomes slippery as I descend beside a stream swollen and noisy with
snowmelt, as cold as steel on the tongue. A couple of grey
moor ponies watch me out of their black eyes. In a bare
thorn tree high on a bank a pair of ravens sits grunting
and canoodling, the male nibbling the female's beak and
having the nape of his neck scratched in return. Eventually
the male spreads his broad wings and sails down over my
head with a sharp interrogative *crrank?* I can hear a musical crash of falling water now, and soon I'm standing on
the rocks at the lip of Light Spout waterfall, a fifteen-footer
that twinkles downwards all in a rush into a vigorously
gushing little rock basin at the foot. I scramble down the
rocky side of the fall and follow the twisting path beside
the stream, on down between steepening hillsides to the
bottom of the hollow. Here the Light Spout stream enters
Carding Mill Valley and turns east for the final mile into
Church Stretton.

Some time during the morning of 30 January, the
snowblind Donald Carr wandered into the top of Light
Spout Hollow. Had he fallen over the main waterfall he
would probably have been killed, but the noise of the
stream under the snow alerted him just in time. Thinking

he was certain to die, and wanting his body to be found as soon as possible so as not to keep his loved ones in suspense, he made one last effort to get to high ground, and in doing so he managed to blunder into Carding Mill Valley.

'I was struggling in a part where the drifts were nearly to my neck, when I heard what I had thought never to hear again – the blessed sound of human voices, children's voices, talking and laughing.' What was the wretched man's despair when his cries for help resulted in a deathly silence. All the children had fled in terror at the sight of him – understandably, as he afterwards acknowledged. 'Doubtless the head of a man protruding from a deep snow drift, crowned and bearded with ice like a ghastly emblem of winter, was a sight to cause panic among children, and one cannot wonder that they ran off to communicate the news that "there was the bogie in the snow".'

Mr Carr's ordeal was almost over. He made it unaided as far as the milling hamlet just down the valley, and the horrified folk there rushed to assist him. Someone helped him to walk down to the Crown Hotel in Church Stretton, where the landlord lent him some dry clothes. 'The effect must have been very ludicrous,' Carr wrote wryly, 'for he was a much stouter man than I was at any time, and now I had shrunk away to nothing.' Then the rector journeyed through the snowbound lanes in a fly, then on horseback, and lastly on Shanks's pony once more, to his own house at Woolstaston. 'As may be supposed there was great rejoicing. So completely had all hope of my safety been given up, that to my people it seemed almost like a resurrection from the dead.'

Very luckily Mr Carr was advised not to dip his deep-frozen hands and feet into hot water, so he kept all his fingers and toes. In fact, after the hundreds of gorse prickles he had unconsciously picked up had worked their way out of his hands, legs and feet, he seems to have suffered no permanent damage from his incredible ordeal. Reflecting in *A Night in the Snow* on the emotion of the moment he finally reached the mill cottages in Carding Mill Valley, Donald Carr gave thanks for his deliverance humbly, as a good pastor should: 'I was saved from the lonely death of horror against which I had wrestled so many hours in mortal conflict, and scarcely knew how to believe that I was once more among my fellow-men, under a kindly, hospitable roof. God's hand had led me thither. No wisdom or power of my own could have availed for my deliverance, when once my sight was so much gone. The Good Shepherd had literally, in very deed, led the blind in a way that he knew not to a refuge of safety and peace.'

I follow in the Revd Carr's footsteps down to Church Stretton, just in time to catch the southbound train among a crowd of home-going schoolchildren. As they tease and text, I watch the peaks and ridges of the Shropshire hills beginning to whiten again with a fresh sprinkle of snow. I watch them, but I don't really see them. I am back in the skull cinema, thinking of Dad and of Donald Carr. The bond of shared values – perseverance, strength in adversity, an adamantine belief in doing one's duty. A throwaway modesty about one's own achievements. The spirit of never-say-die. I nod at my reflection in the carriage window and the distant hills beyond, and I catch myself smiling as I picture my father out on the snowy Long Mynd, relishing a good long walk with the Rector of Woolstaston.

March

The man of March, he sees the Spring
and wonders what the year will bring,
And hopes for better weather.

THE ICE HAS RELAXED its grip on the ditches and water-courses of north Somerset. The waterfall in Scadden's Lane is released to gush and gleam among its cresses. Daffodils are struggling out along the brook, but most of the buds are still hard and waxy to the touch this early in March. 'Look,' says Jane, pointing over a five-barred gate. Two mistle thrushes are pattering in the field beyond, bigger and bulkier birds than their song thrush cousins, but curiously slimline in profile, with strongly speckled breasts and large black eyes. I would have passed the gateway without noticing them, but very little in nature escapes my wife's attention. Even after forty years of walking in company with Jane, my senses aren't nearly as keen as hers. Such as they are, though, I need to rouse them from their winter slumber. Today's climb up the Mendip escarpment is our first proper walk of spring, a preliminary sharpener for a trip later this month to walk the Yorkshire Dales in lambing time.

The mistle thrushes fly away powerfully and in a straight line to an ash tree, where they perch looking conspicuous among the bare twigs as they watch us pass. Stormcock, country people used to call the mistle thrush, because of its propensity to sing loudly in bad weather. But these two are keeping their counsel today. There is high pressure over the UK, and tender weather, chilly but sunny, here in Rodney Stoke at the southern feet of the

Mendip Hills. In the steep field that separates Little Stoke and Big Stoke woods the farmer has spread a reeking load of muck from the cattle shed, and Jane and I have to watch our step among the large, dry, straw-knit lumps as we follow our own steamy exhalations uphill.

On the opposite slope a bare oak stands alone. Its sun-cast shadow on the frosty hillside is a mirror image of the parent tree, as black and perfect as a reflection in still water. Above us the twig tips of the silver birch clumps in Little Stoke Wood are beginning to flush, imparting a wash of milky pink to the sections of the wood where those elegant trees grow. Otherwise all is a dead dun grey.

The path passes through Big Stoke Wood, ancient woodland of ash and small-leaved lime, fairly recently coppiced. Scarlet elf cup fungi, cold and rubbery, lie under the trees like fruit peel that someone has carelessly chucked away. The vividness of the colour is like nothing else in nature at this season, a startling contrast to the frosted grey and brown of the dead leaves that carpet the wood.

At the top of the wood the trees give way to high country. Grey ribs of limestone poke through a thin skin of grass. The rocks ring hollowly under our boots as though giving echo to the caves and passages that form the heart of Mendip, raddled out by all the millions of years of rain-water that have fallen on these hills since they broke clear of the sea. Up at the crest we lean on a gate. A faint haze softens the view – the tower on Glastonbury Tor just discernible, Brent Knoll a surfacing amphibian in the watery levels away west, Quantock and Exmoor beyond as pale grey shadows, the Bristol Channel curving towards the outer sea beneath a hint of Welsh hills. The ridge of Polden, the rise of Blackdown, the Somerset Levels with their geometric parallels of hedges and droves and rhynes where winter's floodwater still silvers the pastures.

'Listen,' says Jane, and now I can hear it too – the first lark song of the year, an outpouring in the upper air, my private marker year by year for the cracking open of spring and new life.

It is a time of death, too. Badgers come bumbling and half asleep from their hibernation tunnels in the Somerset woods, and blunder out into the traffic. They lie along the verges, their white stripes filthy with lorry splashes, their bellies distended with gases. Nothing will touch them, not even hungry buzzards and crows. They linger at the roadside in their sweet-and-sour miasma until another lorry knocks them further into the hedge, or the council road-sweeper picks them up.

Frogs are at risk. There are no wallflowers in the ranine ballrooms of romance. The opening notes of spring have stung all the sleepers into a conga of love. They single-mindedly pursue their search for partners across high roads and dual carriageways. Toads are at it, too, with just as much gusto as their froggy cousins. They teem recklessly out of the ponds and ditches along the old Roman road from Bristol to Wells. Randy toads and frogs with reproduction on their minds are run down and flattened by the dozen, martyrs of love on the B3134. Driving home at night in a storm of sleet we see that the seasonal warning signs, a toad in a red triangle, have been put up along the road. Just beyond, a flash of yellow catches my eye among the silver sleet sparks that slant across the headlights. It's one of Bristol Zoo's toad patrol volunteers in a high-visibility jacket, scouring the verges, bucket in hand, for bulgy-eyed sex addicts who need a lift across the road.

March is drawing to a close as we head north for the Yorkshire Dales by a circuitous route. Clifton-upon-Teme lies in the red-earth country where Worcestershire marches with Herefordshire. The farmer has ploughed knee-deep furrows in the fields behind St Kenelm's church, making stumbly ground to walk on. We can see which fields are habitually wet by the colour of their soil, a rich dark red-brown the colour of drinking chocolate, in contrast to the earth exposed in the drier land which is pale pink, parched and crumbly in the hand. In the steep jungly cleft of Witchery Hole, dog's mercury with its tiny green tassels has taken over the world for a brief moment of glory in the understorey before 'proper' flowers shove it back into the shadows for another eleven months. Bluebell shoots are spearing up. Opposite-leaved golden saxifrage spills tiny flat buttons of yellow-green along the stream. The hazel branches that overhang the water retain the crisp brown curls of last year's leaves. The spent catkins, stripped of their load of pollen by the wind, now hang limp, and the first hard tear-shaped leaf buds are showing at the twig ends.

We cross the bronze-brown River Teme, on whose

bank a split and fallen willow has been parasitized by a couple of dozen bunches of bright acid-green mistletoe, the sharpest colour anywhere in this still-wintry landscape. Climbing to the ridge beyond the river we pass a pond full of jelly streaks, a clotted mass of freshly laid frogspawn. In the wood on Cockshot Hill a pair of treecreepers scuttles up the trunk of a tall, slim ash with a flicker of very white breasts, their heads tucked down between their shoulders, giving them a furtive appearance as they search the bark cracks for hibernating insects. As we watch them, we catch a familiar two-toned song, a crisp and confident *tsip, tsap, tsip, tsap*. It's the first chiffchaff of the year, fresh from completing the small annual miracle of flying from its winter quarters in West Africa to this particular corner of English countryside. Under the leafless trees one or two wood anemones are considering their options. Celandines, primroses and violets are starting to put their heads above the parapet, while in the verge of the lane down to the river a few snowdrop survivors, grey with road spatter, hang their heads and cling doggedly to the last moments of life.

In this upward segment of the year it is all change, in a thousand tiny elements, for those, like Jane, with eyes to see.

I think of springtime at The Leigh, and am amazed at how few details of nature come back to me. The neglected orchards around the village, the dense hedgerows and wildflower meadows were bursting with wildlife back then, but I didn't look at them with particular wonder or a questioning mind. For Roo and me there were two seasons – flood time, and the rest of the year. We took jam jars to collect frogspawn from the pond across the main road – a daring expedition at seven years old. Celandines starred the banks of the derelict cider orchards along the Green Lane as we swished through the puddles, and when the floods had receded from the Big Meadow the marsh marigolds came out to fill its wet hollows with gold. But that was all matter-of-fact stuff. I didn't marvel at the natural world, or observe it with love and fascination – I just ran about in it. Nature came a long way second behind having fun with Roo, breaking fences and sailing hopeless boats. Being outside was all about being naughty boys, running and yelling and falling out of trees.

The chest of drawers in Dad's dressing room at Hoefield House was an irresistible magnet to me, especially after I turned ten years old and began to think that my father must be a spy. James Bond came over the skyline, and that was that. Bond possessed 'dark, rather cruel good looks'. What dark, cruel good looks were like I couldn't really picture, but Dad had black hair and he could get pretty cross. His eyes were blue-grey, too, just like 007's. Fair enough, he didn't have a three-inch scar on his right cheek, and I'd never seen him order a vodka martini or do a racing change in the Morris 8. The idea of him making love with rather cold passion to three married women in turn wasn't a runner either. Obviously he *wasn't* Bond. But I still had no idea what work he did, and he still wouldn't say a word about it. 'Spy' seemed a reasonable bet.

Rummaging in Dad's chest of drawers among his socks and stud boxes, looking for his .25 Beretta, I found a number of double-sided blades, bendy but incredibly sharp as I soon discovered, carefully stored in pale blue paper packets, very small, marked 'Gillette Blue Blades'. The packets had been signed by royalty in a florid hand, 'King C. Gillette'. King Gillette wore a giant soup-strainer moustache in the photo on the blue paper packets, so he hadn't been much good at shaving. I knew that men who'd won medals in the war sometimes had letters after their name, DSO or DFC, or VC for very brave people. King Gillette hadn't won the VC, but he had got a lot of other mysterious letters after his name: 'T.M. REG. U.S. PAT. OFF'. There was no interpreting this cipher from the grown-up world, but King Gillette was evidently an important man. His products in their concealment at the back of the sock drawer had a heady whiff of secrecy and a sliver of danger about them.

My father had letters after his name, too, and some in front: 'Lt-Cdr J. A. F. Somerville, Esq., R.N. (Rtd)'. I would scrawl it all out laboriously, week after week, on the envelopes in which my letters went home from boarding school. Later in life Dad gave short shrift to people who hung on too long to their wartime ranks in civvy street, 'Sqn-Ldr Squiffy Buckmaster' and the like. But back then, ten years or so after the war, those service titles seemed to my generation of children an integral part of a man's identity. Anyway, Dad had been a career sailor, joining the Royal Navy at the age of seventeen as soon as he'd passed out from Dartmouth. And something in the clipped compression of those honorifics, precise yet opaque, was reminiscent of grown-up men like Dad himself, at a time when I didn't see him for months on end, when even in the school holidays he was liable to disappear at no notice for days or weeks on office business. From time to time the official car from GCHQ would scrunch up the drive, the driver would hop out and open the rear door, and Dad would duck inside and be gone.

My obsession with Dad's career as a spy didn't last long. He didn't come home all bandaged up, as Bond did at the conclusion of each adventure. He didn't act flashily, or cheat, or lie. In fact, he was the very pattern of high-minded probity, of self-restraint and duty before all. And he wasn't alone in that. Some men came home and turned their backs on the war that had swept them up. Others, men like Dad, ex-signals officers with good brains, couldn't do likewise. As soon as their war with Germany and Japan was over they were embroiled in a new one, undeclared but intense, with Russia and the Soviet Bloc. Their innate sense of duty, carefully honed and nurtured by their upbringing and education, was tremendously exacerbated by the fact of having survived the war when so many of their fellows did not. They exchanged their officers' pips and stripes for suit, tie and bowler, and they went in where the Cold War was coldest – not in the sewers of East Berlin with .25 Beretta and poison pill, but in back rooms in Hampshire and London with earphones, listening in behind the Iron Curtain.

'Daddy, what do you do at the office all day?'

'I work at my desk and I talk on the telephone.'

'Who do you talk to, Daddy?'

'I talk to some men in another country. It's all very boring.'

I spent my entire childhood in total ignorance of what my father did for a living. It was not until I was eighteen years old, when a guest at a party responded to the phrase 'GCHQ' with 'Ah, the code crackers!' that realization dawned. And it's only now, putting the pieces together ten years after his death with help from those of his surviving colleagues who are willing to talk, that I know about the flights to Gibraltar in the early 1950s to debrief Soviet defectors, the years spent visiting listening posts all round the world, the period in the perilous mid-1960s when he was J, Head of the Soviet Bloc section at GCHQ. 'John's job as J would have been to make sure that other people were doing a good job,' remarked one of those colleagues, still cagey after all these years, '– and he did.'

As the shy and cerebral son of Admiral of the Fleet Sir James Somerville, war leader and famous public figure, Dad knew all about expectations and the impossibility of living up to them. But by God, he tried his level best. He'd stuck it out in the Royal Navy while his father was alive, but the Admiral's death in 1949 freed him to leave the Navy and build a life for himself on his own terms. In the springtime of hope after the war, young parents wanted to make everything perfect, to get everything right – and that included setting a shining, in fact an unattainable example of good manners and sensible behaviour. If I picture Dad striding down the village lane to church, as I so often do, dressed for praying purposes in suit, tie and gleaming shoes, it's as a 1950s icon of responsible fatherhood.

Fathers were like that. If they had troubles of their own, they didn't let on. If they had wartime horrors stashed away, they kept them under lock and key. They didn't give themselves away with tales of their own juvenile misdemeanours. Their love was strong and steady, but it was not explicit. They weren't a boy's best friend or his fun buddy. They were there to set an example and to correct your backslidings. Fathers didn't make mistakes. They knew what to do. They showed you how to dope the tissue wings of a model glider and paint a bedside cupboard with smelly green gloss. They gave you a florin if you cleaned the car properly with a chamois leather, they spoke sternly to you about your school report, and they chastised you if you hit your sister or cheeked your mother. They were upright and dutiful, the object of everyone's respect and admiration. They set the moral bar so high it daunted you.

When I was a baby, Dad was a hands-on father, quite unusually so for a man of his generation. He changed my nappy, he cleaned me up, he sang to me and rocked me in the middle of the night. His love found practical expression, as it did later in his life with his financial planning for his children, his tax advice and injections of funds when I was struggling in the toils of poorly paid teaching. But the Cold War 'civil servant' of the 1950s with his freight of international responsibilities and his private anxieties, coming home each evening unable to speak a word about any of it, even to his wife – that version of Daddy was not amused by stories of the naughty words I'd uttered or the apple branches I'd broken that day.

Dad's main escape from work and family pressure was walking. He wanted and he needed to stride many miles, alone if he could, rain or shine, in an old leather-elbowed coat, tattered trousers and fawn mac with two missing buttons. Everything about his preparations was very sparse and frugal. No pies, no pints, no pubs – God forbid. That sort of thing was for self-indulgent chaps who were busy 'digging their graves with their teeth' – a favourite expression of Dad's. Into his blue-grey webbing haversack went a soapy slice of 'fridge-aged' Cheddar between two Ryvitas, a plastic bottle of weak lemonade and a withered apple from the attic floor where he stored dozens of pippins each autumn on newspaper beds. He'd consult the cloth-backed one-inch Ordnance Survey map with its minuscule, all-but-indecipherable dots showing the footpaths that might or might not be there on the ground. The right-angled ash walking stick would be drawn out of the hall stand, and he would be off to the nearby Cotswolds or over to the Brecon Beacons for a dozen miles at a good clinking pace, walking it all out of his system, to return home muddy and rain-spattered and at peace with himself for the next few hours at least.

I think of my own walking gear nowadays, the padded Páramo coat that sheds the rain and bars the wind and wicks the sweat away, the wonderful boots with their high-tech soles that steady me over bog and rock, the thermals that keep me warm, the gaiters that absorb the mud, the Satmap device that won't let me go wrong (if I've remembered to plug it in), and I blush for my lack of hardihood and the cushioned experience I have when I'm out on the hill. Rain, hail, snow, wind, mud – they all got through to Dad, and he strode through the lot of them like a lord.

A shift in space and season now, in the last days of March – a hundred and fifty miles north from the awakening red-earth country of Worcestershire to the gritstone moors of West Yorkshire where spring is still no more than a rumour. Jane installs herself in the warmth of a tearoom down in Hebden Bridge, and I climb to Horsehold Farm up a lane whose green and slippery cobbles are coated with frosted algae. The moors above, exhausted by two months of snow, are blanketed in brown and cream. It's whirling weather up here, quick-change weather that puts me all at sixes and sevens. One minute a smack of warmth from a clear sun in a blue sky has me stripping off my heavy winter layers; the next moment an icy blast from the north comes blustering across the hills, and it's on with the scarf and gloves and up with the hood. Grey murk rushes past, one field away, leaving me untouched. A rainbow frames the mill chimneys of Mytholm in their chink of dale bottom four hundred feet below. And a shower of sleet and thin snow blows through Callis Wood, driving me back down the track through this witchy place with its black hollows where winter still skulks, where the bare boughs of silver birch and oak reach across the path like sinewy arms that would hold back the spring if they could.

The long slim valley of Nidderdale winds and wriggles north-west from Harrogate for twenty miles, narrowing all the time and climbing until it turns west and pushes its snakehead into Scar House Reservoir under North Moor. Up there a ewe lies on her side in the corner of a field, in the angle of a gritstone wall that sparkles with mica. Her first lamb has just been born. Its wool bright yellow with amniotic fluid, it is struggling up onto its legs beside her, a red worm of umbilical cord dangling beneath its belly. The ewe lies on her right side, her face expressionless, her body quivering with the effort of giving birth to the second of her twins. Slowly it begins to emerge, a vacuum-packed lamb in a silvery, glistening membrane bag. The ewe heaves herself up to stand and let gravity do the rest, and the lamb drops out onto the grass with a slithering thump, the membrane stretching alongside, crimson with rich blood.

The newborn lamb lies still, its outsize ears flattened to a head crooked back so far it seems as though its neck is broken. With the vivid orange-yellow of its skin barred by dark lines of wrinkles, it looks more like a huge wasp or a tiny tiger than a lamb. The mother turns and licks up the membrane, eating it off the lamb whose umbilical cord has already snapped. Its older twin has got itself up on its legs. It watches the scene for a minute, then turns and totters under the ewe towards her udder. Within a minute or so the newborn gets its head up and nuzzles its mother's face. The long legs twitch, and the nuzzling continues along the mother's neck and body, as the lamb inches its head back towards the smell of milk. The ewe licks and licks, cleaning the gunk off the newborn, and now the older lamb disengages its mouth from her udder and gives the head of its twin a lick, too.

The newborn gets up jerkily, wobbling on legs that are lumpy and seem disarticulated. It staggers back towards the udder, burrows under the dung-clotted fringe of its mother's fleece, and finally gets its mouth on the other teat, sucking tentatively, then vigorously, while the ewe licks and cleans the loosely wrinkled skin of amniotic orange.

This is the month when shepherds get scant sleep. All up Nidderdale, in warm sheds and out under stone walls and alongside barbed wire fences, ewes are giving birth. The grey racing sky and the green slopes of the dale are loud with cries. The tremulous calls of tiny lambs and the gruff, phlegmy responses of their mothers are clearly heard down among the dark stone houses of Nidderdale's scattered hamlets. On a blowy morning Jane and I set out from the dale-bottom town of Pateley Bridge along the Nidderdale Way to reach the lonely cleft of Ashfold Side.

Beyond the tightly huddled houses of the town rises the 'inbye land', the daleside fields nearest to the farmhouses, squiggled with the black lines of stone walls. These have the appearance of wriggling like snakes as they race across the undulations of the ground and vanish over the skyline. This is beautiful farming country, but hard as nails in winter, the farmhouses and barns built all of a piece, the classic Dales longhouse reflecting the architecture that the Danes brought with them when they came to Nidderdale in the mid-ninth century with fire, sword and a hunger for land.

We follow a lane flanked by sentinel beeches with silvery bark as finely wrinkled as a nonagenarian's skin. Their tops, still leafless, resemble puffs of thin grey smoke. The black and green walls along the lane are built of big bouldery stones, knitted together with long 'through' stones. They are roofed with smaller, roughly shaped capstones, set on a layer of single split stones to weatherproof the main body of the wall against rain dripping between the capstones. One section next to a gateway, bashed down by a farm trailer, has been repaired by local offenders – a neat and useful way of paying back the community for their petty thieving and litter-louting. The structure of these gritstone walls has a pleasing simplicity and fitness for purpose, but they cost a lot of time and effort to keep up. We pass a man rebuilding a section. He bends and straightens, his round face crimson with exertion as he throws the tumbled stones from one pile to another, his palms protected by big industrial gauntlets. 'Quite a job,' he grins, straightening up and blowing hard. 'If you've got half an hour, you're welcome to tek over.'

The fields are full of sheep. Most are Swaledales with black faces and white muzzles, but near Bale Bank Farm we spot a more exotic couple of Lincoln Longwools with lumpy rasta dreads covering their eyes and a most endearing silly grin. Beyond Low Hole Bottom a gang of twenty Swaledales comes up, bleating plaintively as we sit on a fallen stone gatepost eating ginger oatcakes. They are mendicants, their palates ruined by the chocolate and crisps of picnickers on the Nidderdale Way. Each ewe wears a pair of eartags, one orange, the other yellow. A couple are limping from foot-rot, and one is shuffling on her front knees to relieve the pain. Another of the gang has lost her whole front foot, and goes awkwardly along the hard stony track, dot-and-carry-one, favouring the truncated leg, every step an obvious agony. It's a gait so painful-looking that I can't bear to watch. A human would be wincing, gasping, screwing up eyes and mouth. But all the ewe's suffering is masked behind her immobile face.

The air from the moors is cold enough to prickle the nostrils, smelling of earth and leaf rot. Big silver and grey clouds sail twenty thousand feet tall, blown by a gale of north-west wind in the upper air. Closer to earth, however, it's almost completely still. They are burning the heather up on the grouse moors, and the columns of oily blue-and-gold smoke rise vertically into the air with hardly a waver.

A hoarse rattling draws my binoculars to the rim of the dale. An incautious red grouse is standing silhouetted on a rock, rising on tiptoes to shake his wings. Behind Bale Bank Farm a curlew launches its pulsating bubble of a cry, the indescribably sweet and poignant call that will haunt all these sedgy uplands for the next few months as the birds settle in for the breeding season. A lapwing goes tumbling over the fields in a courtship display to a potential mate wheeling in circles below, crying *Piddy-wee! Wee-wee! Peek, piddy-peek!* A black-headed gull already in summer plumage perches on a fence post, its dark head blending so neatly with the dark surroundings that it appears headless, the light grey wings and white body rising to a sharp point with nothing on top, a disconcerting sight.

Beyond humpbacked Brandstone Dub Bridge we enter a no-man's-land between the green inbye and the brown moor. Old heather bushes, tough and blackened, have been left as shelter for the grouse. Beside the track Jane stoops and calls my attention to the bones of a rabbit packed into a chink below some raptor's chopping stone. The ribs curve like those of a tiny boat. The delicate vertebrae are hollow and winged, the leg bones long and pale green with algae.

The spoil heaps of Prosperous Mine appear ahead, at first disguised as a rubble of loose stones poking through the coarse grass, then naked and blatant, a mass of large pale grey banks of barren waste that slope in frozen motion down the valley of Ashfold Side Beck. A pair of grouse stands on top of one spoil bank, the male, with his magnificent red eyebrows, bowing and screeching his mating dirge to a drab brown and utterly indifferent female. These Mordor-like leavings of the lead-mining industry, ashy and drained of colour and life, make a stark contrast with the semi-wild beauty of the countryside they scar, the deep rocky cleft of Ashfold Side, the rustling beck, the green dalesides with their ferns and trees. Yet they are inextricably linked with its story.

The Prosperous Mine had its heyday in the nineteenth century, but this area had been mined for lead since Roman times. Prosperous in its remote side dale was certainly being worked in 1780, the ore being smelted and processed on site. It was an off-and-on kind of set-up, closing when lead prices were low, opening up again when they improved, till the final death knell in 1889. The full title of the outfit, the Prosperous and Providence Lead Mining Company, smacks of starry-eyed optimism tinged with wishful thinking, as do the names it gave to its mining levels above Ashfold Side Beck – the Wonderful and the Perseverance. Cornishmen, Irishmen, Scots and Welsh worked here alongside the men of Nidderdale. The clash and mingling of accents and cultures must have been something to behold. The workers' living conditions out on the moor were harsh in the extreme, their continued employment precarious, their health constantly at risk from a habitual diet of bread and tea and from the poisonous fumes associated with lead smelting, and their lives likely to be ruined or cut short by industrial accidents.

High over Ashfold Side Beck we find a tall pyramid of spoil, and a narrow, flimsily fenced-off mine shaft – a triangle of rusty girders over the mouth, a dark funnel chuting away straight down into blackness. Fragments of buildings lie smothered under spoil near the dressing floors where the ore-bearing rock was broken up. Down by the beck stand the ruined walls of the water-powered smelt mill, with a handsome arch held up by sketchy timber framing. A waterwheel drove a bellows for the smelting hearth here – the rusty cogwheel of the mine pump, horizontal on its shaft, still stands upstream of the mill ruin. Stone-lined flues took the fumes away to a venting chimney up the bank. Rusty machine parts, enigmatic levers and boiler plates lie all around, their function as obsolete as a broadsword unearthed from a bog or a horse drill forgotten in a hedge. Ashfold Side Beck runs mineral orange between banks scattered with pale drifts of spoil, until a mile downstream it shudders over a low fall and shakes off the trammels of industry in a sparkling run of ice-clear water.

The pastures around Highfield on the way over to Nidderdale are full of tiny lambs trotting after their mothers. Each lamb's tail is tightly bound with a red rubber band an inch from the root. The appendage below the ring will rot through lack of circulation, shrivel up and fall off within a week or so, leaving the lamb with cleaner hindquarters that are less liable to be eaten away by maggots. Fly strike, as it's known, is no joke for a sheep, or the shepherd who has to deal with it.

In a field beside the lane a wild mallard drake patrols a stream in company with a white domestic duck. A little distance away a female mallard watches the pair, quacking discontentedly as though to say: 'What's she got that I haven't?' Below the lane on the steep grass slope down to Stripe Head Farm, dead moles lie beside their holes, each corpse hideously contorted, the black velvet bodies crisp and dried, their great white claws snapped shut in their palms. What has killed them? Poison? An aggressor, perhaps a fox or a stoat? A terrier? But don't moles taste nasty to dogs? I have seen dogs, and cats too, sneeze and recoil in disgust from dead moles on my mother's lawn. So what would kill a mole and shake it inside out like this? For the hundredth time this year, it's borne in on me how very little I know about everything I see.

Down at Stripe Head Farm, big strong Texel ewes are lambing. The breed was introduced into this country in the 1970s from the Dutch Frisian island of Texel, and it's been a great success, able to do well even on thin upland feeding, and really thriving on the richer grass down in the inbye land. As we cross the field next to the farmyard we notice a Texel ewe brusquely shoving two tiny lambs away with her broad, blunt head. She emits loud, blaring calls of distress. The gate clicks, and the farmer in gumboots and flat cap comes into the pasture to see what the matter is. He picks up the lambs one after the other by the loose scruffs of their necks to inspect them.

'Them lambs was born this afternoon, maybe a couple of hours ago. She might have rejected one because it's been lying with another ewe's lambs and smells of them. Or maybe one of the quarters of her udder's not working – I'll have to check that.

'I've a flock of two hundred lambing just now. We've been getting a lot of triplets because last year's grass was so good. We started last week down here, but up on the dale top they'll be a week or two behind. They're a few hundred feet higher, that's why. The gimmers, that's the young ewes, usually like to lamb on their own. They'll find a corner or go behind a wall – to be private, like. Sometimes we'll see one with something hanging out of her back end and we'll know she's about to give birth. I'll bring her inside then. It gets pretty crowded – they don't take up much room when you see them on a fellside, but once you've got them indoors they take up a lot of space sideways! I'll bring her in to disrupt her timetable, like. She thinks, well, he must have got me in here for something. So I'd better get on wi' it!

'Some farmers get up and stay up all night at lambing, but not me. I go out at 11 at night to check on them, and I'm out at 5.30 a.m. again. I'll know if anything's up with one. They *let* you know, you see.'

On a freezing cold morning we walk the narrow road that runs north up Nidderdale, a snake of a lane between stone walls. An obsidian glint between the willows shows where the River Nidd flows, a field's width away from the road. Skeleton trees stand very black against the green of the dale sides. Under the dark stone field walls, ewes heavy with unborn lambs lie panting. We pass the dam at the southern end of Gouthwaite Reservoir, its walls and lodge embellished in baronial gothic style. A shepherd putters towards us on a quad, hay bale on the front, sheepdog on the back, head down, beanie hat clamped on head, cheeks pulled back by the wind of his passage into a manic grin.

Where the dale road swings next to the river, Jane spots a flash of white, all but concealed in the roots of a willow. It's a mallard's nest the size of a soup dish, constructed of tiny pieces of chopped-up twig lined with soft downy breast feathers. There is a pile of large oval eggs inside, fifteen of them. The mother has only just left the nest. I'm on my knees photographing the eggs when Jane hisses 'Kingfisher!', and there he is, darting across the river to perch on a low twig, his tangerine breast glowing in the frosty light, his cerulean back speckled with points of sparkling electric blue. I can't work the camera quickly or well enough to catch him; and anyway, committing the moment to memory rather than to a blur on an electronic chip seems a better way to spend the few seconds we are in his presence.

The deep little gorge of How Stean, only a couple of armspans across at its narrowest, branches off north-west from the main dale at Lofthouse. How Stean has been equipped for outdoor adventure with high wires, walkways and perilous bridges. High-pitched shrieking echoes up from the gorge bottom, where a party of excited children in helmets and overalls is negotiating the rocks, shallows and rapids. The overhanging rock walls sprout ferns and mosses. Streamlets trickle over slick mats of shining green liverworts. In the depths How Stean Beck rushes over slabs and swirls round churn holes and through black narrows.

The breath of the gorge is cold and misty on this March afternoon. I pay my fee and embark on a hand-railed path near the rim. Soon a flight of slippery stone steps leaves the path and descends almost to river level. Down there a trail runs parallel with the path above, an exciting scramble along the gorge walls as they swing and snake through the limestone. I start off along the catwalk, guiding myself by a hand-wire fixed in long loops to the wall on my left. At a certain point, just where a couple of cast-iron lizards are riveted to the gorge wall, I feel my walking boots skid from under me on the slippery ledge. My knees knock against rock that's slick with algae, and I cling rather desperately to the wire with my left hand and the nearest lizard with the right, while I scrabble for a foothold I can't feel. It's an uncomfortable half-minute hanging by my arms above the rocks and rushing river potholes twenty feet below, and then I find some purchase and manage to haul myself up. My heart bangs like a trip-hammer. I release my grip of the wire and find it has left a red welt across my left hand. The lizard's cast-iron skin pattern is similarly imprinted on the other palm. I am very glad to crawl along to the next set of steps and up to the main trail again. Looking down at the scene, I realize I have been lucky. If I'd fallen, I'd certainly have bashed my helmetless head on the rocks and slid stunned into several feet of icy, racing water.

Back up the gorge a rickety bridge crosses high above the beck and leads to the mouth of Tom Taylor's Tunnel, a natural passage running north through a cave in the rock to emerge in a nearby field. Tom Taylor, the information board tells me, was an eighteenth-century outlaw, but not a romantic one. He was a nasty piece of work, a member of a criminal gang who would break into farmhouses and torture the inhabitants till they revealed where their valuables were hidden. After a while Tom quit the gang and went solo as a highwayman, maintaining a hideout in this crack in the wall of How Stean Gorge. But someone gave the place away, and a lynch party set out to capture the outlaw. Tom heard them coming, and managed to squeeze himself up onto a ledge in the roof as they entered the cave. There he commenced hooting like an owl. The vigilantes, convinced that no owl would have stayed put if a human had been anywhere near it, left the cave. But a few days later they returned. This time Tom was minus his front teeth – they'd been knocked out in a scuffle with a constable. He climbed to his ledge and tried the same trick as before. But the gap in his gnashers distorted his owl impression. The unconvincing hooting gave him away, and he was seized and summarily hanged from the cave roof.

I climb down a ladder to the start of Tom Taylor's Tunnel and shuffle along the triangular cleft, torch in hand. The boat-shaped roof pulses and ripples in the wavering light of the torch like a laparoscopy film. Stalactites drip on my head, devil's-toenail fossils gleam in the walls at eye level, and there's a shivering chill in the dead-still air of the cave. It's a spooky place. I don't pursue the tunnel to its outlet in the field, but turn tail and make for the bridge and the upper world where a low sun is beginning to melt the frost along the river.

In the taproom of the Crown Hotel, Lofthouse:
Landlord (greeting regular): 'Alan.'

Customer: 'Mm.'
Landlord: 'Theakston?'
Customer: 'Aye.'
Landlord: 'Pint?'
Customer: 'Aye.'
A minute passes.
Landlord (placing full glass before customer): 'Now.'
Customer: 'Ta.'
A minute passes.
Landlord (reaching for empty glass): 'Mm?'
Customer: 'Mm.'
A minute passes.
Landlord (placing refilled glass before customer): 'Now.'
Customer: 'Ta.'

A blue sky above Nidderdale on one of those rare peerless evenings on the threshold of spring. We sit in the window of a barn above Wath and look out up the dale. In the next field, lambs born three or four days ago are careering around in a posse eight or ten strong. They gallop madly across the field, stop at the far fence to sniff the air and each other, and then one of them kicks back across the field and the rest come dashing behind. Solitaries are swept up and carried along in the river of lambs as it pours past. They leap in the air, all four hooves off the ground, as if trying to kick an itch away. They are still occasionally wobbly on their legs, and every now and then one goes staggering sideways without warning, taking itself by surprise. Their skins are wrinkled and look too big for them, as though they are wearing an older sibling's hand-me-down onesie. As dusk falls their play becomes less skittish. They suddenly collapse and sleep where they fall, some by the stone walls, others in the shelter of their mothers' big, bulwark-like bodies.

A pair of twins is trying to get a last drink. One has four black knee socks, the other a narrow black strip like a silk stocking seam down the back of each leg. They come in under their mothers' shaggy pelmet of wool and butt their heads up hard to stimulate the milk flow. She tolerates it for a minute or two and then moves forward, unplugging the two mouths. The lambs scamper to get in front of her and try to push her to a standstill with their heads, like two tiny tugs trying to shove a tanker into reverse. She comes to a stop in her own good time, and continues eating very selectively and delicately, concentrating on one tiny patch of ground and pulling up her chosen mouthfuls with short, decisive downward tugs of her teeth.

Eventually the whole field settles. The ewe flops down, and her blackleg twins lie in shelter to leeward of her. Two lambs by the wall get up and scamper off to a ewe who turns out not to be their mother and has her own twins to feed. She butts and plunges at the interlopers, then waddles crossly away. They give chase in forlorn hope of a nightcap. Soon all the lambs are paired with their mothers. The grunting calls of ewes and baby wails of lambs diminish and stop. A rising wind sends a few shreds of smoky cloud across Nidderdale from the north-west, and

the blue of the sky deepens and darkens. There will be a
sharp frost tonight, and in the morning we will walk out
to find every lamb sporting a coat of sparkles.

April

Through April rain the man goes down to
watch the birds come in to share the Summer . . .

EASTER SUNDAY MORNING IN our village church. Ralph Boddy has laid a bed of fresh moss on the sill under the east window, and has built on it the simple message 'HE IS RISEN'. The letters are composed of a hundred and twenty bunches of primroses that Ralph has picked especially for the day, just as his grandfather and three older generations of the family have done before him.

This annual gift of Ralph's to the village is totally dependent on the season. Some years the primroses are almost over by Easter; other years they have hardly budded. But Ralph is a sort of magician. Year after year without fail, no matter how early or late the primroses, he finds enough to form the nine letters of his message. He doesn't particularly want to reveal his sources. 'I've got my places,' is all he'll say on the matter.

This Easter morning the message on the east window-sill brings pain as well as pleasure. I have just had word from Crete that my dear friend Lambros Papoutsakis has lost his wife Maria. If it is possible to die of a broken heart, that's what Maria Papoutsakis has done, five years after the hang-gliding accident that killed her eldest son, Giorgos. 'Since Giorgos gone, never she laughed again,' Lambros says in his message. I remember wishing, 'Christos Anesti, Christ is Risen,' to Lambros, Maria and Giorgos, and their ritual response: 'Alethos Anesti, Truly Risen.' It is certainly possible to feel an aching heart.

*

All of a sudden, two weeks after Easter, April becomes the yellow month. The countryside is flooded with yellow – the oily yellow of celandine stars, soft sherbet yellow of primroses, clear Dayglo golden-yellow of dandelions. Wrens and chiffchaffs are in full song. In the south Cotswold woods the mosses are spattered with violets and wood anemones. Jane spots the first tentative bluebell flower, the surface of its emergent petals streaked grey-blue like muscles revealed under flayed skin. Hedges are beginning to prick with green shoots. I realize that I have seen more traditional hedge-laying in the last ten years than in all the rest of my adult life, a real resurgence of the craft, especially in the southern counties. This time last year we passed a beautifully laid hedge in the lane at Lower Kilcott, done by someone with time and care to spare, the stems cut three quarters of the way through near the hedge root and bent at a near horizontal angle. However, this year a mechanical slasher has given the top of the hedge a brutal, splintery crewcut. The contrast between the recent brisk butchery and last year's pains-taking craftsmanship is quite striking.

Two weeks after Easter I head north along the M5. It's cloudy and cold, the car thermometer showing 5°C. Tree flowers are out – hawthorn, blackthorn, prunus. Pussy-willow catkins have exploded. Hawthorn leaves are breaking, the bursting buds thickening the outlines of the trees. The lambs in the Gloucestershire pastures are fat and sturdy, their thighs bulging like all-in wrestlers'. In

Worcestershire the cattle have been let out of their winter sheds into the fields.

Near Stafford I overtake a lorry promising 'Integrated Logistics Solutions'. Didn't that use to be 'Haulage and Storage'? The central reservation flashes by a couple of feet away, a streak of green and white. What's that white component? Some sort of small flower, but I can't spare it more than a glance. Later I look it up, and am amazed. Why haven't I been aware of *Cochlearia danica*, Danish scurvy grass, before today? It's the fastest-spreading plant in the UK, I read, a halophyte (salt-tolerant – I look that up, too) member of the cabbage family, whose sharp-tasting leaves are full of vitamin C. Seafarers on long voyages used to eat it from salt barrels to ward off scurvy. It started off in this country as a coastal plant, but has found motorway travel very much to its liking. Danish scurvy grass thrives in salty conditions on well-drained ground. Motorways are very well drained, and they are salted against ice. An added advantage for *Cochlearia danica* is that the motorway salt burns bare patches along the hard shoulder, rendering this thin strip of no-man's-land inhospitable for rival species. And the small seeds of the plant spread themselves up the motorway by slipstreaming on the bow-waves of air created by vehicles belting along in the fast lane. They surf inland on these salty aeolian waves, colonizing the linear world of the central reservations, a migrant tide that spreads and thrives beneath the radar.

I turn west off the motorway into Cheshire to stretch my legs along the Sandstone Trail. This high-level path hurdles the back of the remarkable sandstone ridge that runs north through the county all the way from the Shropshire border to the River Mersey. The ridge stands three hundred feet above the surrounding plain, a knobby saurian spine leading the eye northward to where Beeston Castle at the edge of its crag forms the head of the beast.

Spring comes forward to meet me in patriotic colours as I follow a woodland path between the ridge and the plain. Red dead-nettle. Sky-blue flowers of the perversely named green alkanet. The first stitchwort of the year in white stars. Damp dells and ditches are brilliant with the incandescent cadmium yellow of kingcups, like drops of gold on the juicy black earth. From the mossy cracks within the sandstone walls of the Peckforton Estate droop porcelain-white bells of wood sorrel, each petal scribbled with fine looping lines of mauve. Ferns in the pink sandstone soil under the oaks are beginning to unfurl their fiddlehead scrolls. On patches of heathy ground further up the slope, bilberry is starting to bloom, the blue-grey calyx cupping the delicate, blush-pink bell of the flower.

The path to Beeston Castle crosses fields of dark soil where the hedge hawthorns are still flowerless. Their many-fingered leaves are fully out, though. On the neighbouring blackthorn, by contrast, the delicate white flowers have opened, but no leaves have yet shown on the black thorny twigs. The ash flowers are out, a dark crimson fletching round the sprouts of unopened leaves. Each bud, a sheath of dark velvet green, is in the process of being split open by the power of what it encloses, a pale green stub of leaves pushing out of confinement, as tight and hard as a finger.

An old collapsed remnant of goat willow lies prone on a bank, its few desiccated limbs reaching out of a white tangle of dry nettle stalks. The structure couldn't look more utterly dead. Yet each horizontal branch has thrust up a rank of vertical young shoots, three feet high and as straight and vigorous as guardsmen, each studded with busbies of pussy-willow catkins. Every one of these soft grey cat's-paws is bristly with anthers tipped with egg-yellow pollen. The old goat is as sappy as a young kid, in spite of appearances.

Everything in nature seems bursting out of bounds today. A *Magnolia grandiflora* on the verge of the lane has scattered its largesse of big white flowers, the petals with their artichoke-shaped bases lying like exotic feathers in the grass. A chaffinch in the hedge does its fast-bowler impression, an accelerating run-up with an explosive follow-through: *chee . . . chee . . . chee-chee cheecheecheechee – PTCHEEUW!* A blue tit forages high in an oak, calling *see-wee!* in a scratchy little voice. Sparrows hop and chatter in the hedges around Outlands Farm. Little bull calves confront each other in the pastures at Wood Farm, pushing their blunt heads close together as though taunting one another, then wheeling round, kicking their hind legs and tails high as they bucket away across the grass.

The first swallow of the year cuts across the field as I walk towards Beeston Castle. I pass through the massive sandstone gatehouse and climb up the slope of the inner ward where I find a timeless frolic in play. A doe rabbit sits motionless on the bank of the moat. A buck scampers close and darts round behind her. He leaps up on her back, grasping her with his forepaws, and thrusts away – ten jerky stabs in about three seconds. A kick from the doe's powerful hind legs sends him sprawling, but not for long; he springs up, and the pair of them go leaping and darting all over the bank, playing lapine kiss-chase – big fat rabbits with golden ochre collars of fur, as healthy as can be, lively and quick in their springtime manoeuvres.

For years I've been promising myself a trip to Marshside in spring. This RSPB nature reserve clings to the lower lip of the River Ribble where it debouches on the Lancashire coast in a wide-mouthed estuary. Marshside is a reserve of two halves. Seaward it comprises a great expanse of salt marsh, a mile wide, leading out to sands that at low tide can extend as much as four miles further west before they meet the edge of the sea. The landward portion of the reserve is formed by a pair of adjacent freshwater marshes – Sutton's Marsh, whose pools and fleets of water contain islets for nesting birds, and Rimmer's Marsh, half golf course, half reedbeds and grazing meadows. In spring the air and sea, the sands, the marshes and pools are brimming with birds, some intent on staying put and breeding, others stopping off to rest and feed on their epic journeys of migration to the roof of the world. Marshside at this time of the year is one of the best spots in these islands for a birdwatching walk. But I've never stirred my stumps sufficiently to get there at the right season. This year, at last, I have the rhyme and the reason. 'Through April rain the man goes down to watch the birds come in to share the Summer . . .' Dave Goulder's line from 'The January Man' gives me the push I need.

Nick Godden, RSPB warden at Marshside, is waiting in the reserve's car park. 'Oh, look,' is his greeting as he raises his binoculars skyward, 'our pair of Mediterranean gulls!' There they are above us, on stiff wings so brilliantly white they look translucent, with big black heads and thick scarlet bills. 'They were mating last weekend, so we think they may be going to nest with us, which would be a first. We're very excited!'

The main hide on Sutton's Marsh overlooks a lagoon. It offers us a grandstand view of an avocet's nest, a thin pile of scraped-together twigs and stones on a tiny man-made shingle islet that can't be more than thirty feet away. The water level is kept low so that the avocets on long blue legs can stalk the shallows of the lagoon and sweep it, as the males are doing now, with their gracefully upturned bills, sifting out worms, insects and tiny crustacea. The trouble with water this shallow is that predators can cross it. Foxes have been known to wade out to the nesting islets. The RSPB has erected a four-foot electric fence right round the marsh to keep them out, but even so they can't be sure that a hungry fox might not decide to brave the shock for the rewards on offer.

'I saw the first avocet eggs being laid last Saturday,' murmurs Nick. He describes the egg growing inside the female. From a single fertilized cell to a full-size egg ready to be laid, the process takes just one day. The female needs a huge intake of calcium to form such a rapidly growing shell. She gets it from the exoskeletons of the invertebrates she eats, tiny crabs and suchlike, and also from the shells she forages for in the evening, grinding them down overnight in her gizzard with bits of stone or gravel she's swallowed. Avocet males and females share nest-sitting duties, and they arrange their changeover so that the eggs aren't left exposed to gulls or foxes, magpies, crows or any of the other chancers who keep their beady eyes wide open on the reserve in the nesting season.

Once more I'm struck by how small touches of detail accumulate to prise open one's eyes and imagination, revealing a rainbow picture created by dozens of monochrome snippets of fact. And I marvel to myself at the availability of experts such as the warden of Marshside, sitting beside me at the hide window, generous with his time and knowledge, patient with my ignorance, prepared to let me tap his stores of information as freely as I like.

Over to our right, black-headed gulls have created an island state on a man-made raft. The gulls are in their summer plumage with chocolate-brown hoods and crimson bills and legs. They are the noisy neighbours of the bird world, forever coming and going, screeching and fussing. Nightclub manners are in order. Male gulls drop beside the female they have their eye on, and regurgitate a dollop of food in front of her. If she accepts it, that's the 'OK'. Get your coat, love, you've pulled.

Across the main Southport highway lie the enormous salt marshes and sands of Marshside's 'other half'. Seaward over the marshes winds the Old Haul Road. Carts full of sand dug from the edge of the salt marsh were hauled inland along this tidal track to a processing plant for the building trade. The RSPB have renamed the track 'Redshank Road', but I prefer the rough poetry of the older name. Nick and I walk the Old Haul Road through a dead-flat landscape. The tide comes creeping up the track from the far end, quietly gurgling in the runnels and creeks at either hand. Skylarks craze the blue sky overhead. The marsh is patched with the loose white flowers of scurvy grass, lying like localized snowfall. I've learned a bit about the plant since *Cochlearia danica* went flashing by on the hard shoulder of the M5 two days ago. I pick one of the leathery, ascorbic leaves, and find it very sharp and tangy in the mouth.

Out at the edge of the sea forty dunlin go flying in a dark cloud close to the water with wings extended, then flickering all together like silver paper as they turn and beat their wings as though they are one entity, displaying their white wing stripes and the black belly patches they sport as courting suits in summer. Allelomimesis, says Nick. What a fine word to learn. It means that clustering mechanism that drives starlings to ball together several million strong, or a thousand flying waders to swerve into a new course, seemingly as one organism.

A redshank stalks the banks of a rapidly filling creek. Seeing our tall shapes approaching, it flies off with short, piercing calls of alarm: *tic, tieu, tieu! ptieu, tieu!* 'The Warden of the Marshes,' says Nick, 'they're the first ones that kick off when any danger becomes apparent.'

Redshank nest in tussocks, pulling the long grasses over to make a dome that's very hard to spot. The dome has another practical use for a bird that nests in a flood-prone habitat. Today is an unusually high spring tide on the Lancashire coast, so many skylark and meadow pipit nests and their egg clutches will be overwhelmed and lost (the parent birds may well have a second brood as compensation). Voles will be washed out of their holes and provide easy pickings for raptors. But redshank eggs can survive at least an hour underwater, and the parent birds won't find them dispersed beyond retrieval once the tide has receded – the grass dome will have kept all the eggs together in one place.

This walk through the marshes of Marshside becomes a humbling eye-opener for me as Nick Godden turns the spotlight of his knowledge on this bird and that. A golden plover emits short, intense cries as it flies overhead. It'll soon be on its way further north to its breeding grounds – probably in Iceland these days, remarks the warden. There are big problems for golden plovers, he says. Craneflies are the plovers' main food source, and the birds time their mating so that their eggs will hatch in May and June, the months when the craneflies are emerging as adults from the peat bogs of the north. But climate change is threatening the golden plover with a population crash. The UK is at the southern edge of their summer or breeding range. As the world warms up, there's a general tendency for that southern edge to shift ever further north to cooler climes – in the case of the golden plover, to Iceland. Hotter summers mean drier surfaces in the bogs, killing off the larvae which produce the following year's adult craneflies. Food shortage results, and a diminished population.

It's not just the warmer summers that are taxing the survival skills of wild birds. In winter, Nick says, many species – goldeneye, goosander and tufted duck, for example – are 'short-stopping', staying closer to their Arctic breeding grounds, because the unfrozen lakes and places where they can find food are shifting further north as the world warms. They don't have to travel as far south as they used to in order to find open water. No duck with any sense is going to fly any further than it needs when the season is cold, food is scarce and energy is in short supply. So around 50 per cent more of these ducks are wintering in Sweden and Finland now than thirty years ago, and the same percentage fewer are continuing south to winter in northern Europe on the latitude of the UK. They are adapting to the change in their circumstances. But how much further north can they go before they run out of breeding space and food? And how many of their numbers can they lose to human depredation before they experience a population crash? They are no longer visiting the more southerly reserves and protected areas set up by conservationists as safe winter havens where everything is provided for their wellbeing. Further north around their new Scandinavian wintering haunts, people are not necessarily as compliant with conservation ethics as folk in the UK are. Guns and traps lie in wait. It's no joke for wild duck in these new, uncertain times.

Back on the inland side of the reserve I say goodbye – and a heartfelt thanks – to Nick Godden, and set out to walk the perimeter paths and roads of Sutton's Marsh and

Rimmer's Marsh. A pair of pintail is sailing on the Junction Pool, the male very beautiful with a white streak down his neck and a stiletto tail almost as long as his body. They are stoking up before their onward journey to Russia where they will mate. I sense again the restlessness of nature, the implicit journeying of these two birds who dabble for their food in a Lancashire lagoon today, and tomorrow may be flying over the frozen Baltic five hundred miles to the east.

It's funny how attached you become to objects, particularly familiar ones that you've painstakingly learned how to work. Lost mobile? A nightmare. Broken computer? Don't even think it. When I mislaid my 'best' binoculars a year or two ago, I mourned them like a lost dog. The ones I replaced them with, a pair of Optolyths, are a bit clunky, a bit clumsy. I don't appreciate the weight round my neck. But I have to admit that they are absolutely the business when it comes to looking through them at stuff. The main area of Sutton's Marsh is a maze of pools, ditches, boggy areas and wet grassland, confusing ground for picking up the well-camouflaged nests of lapwings in their nesting habitat. But the Optolyths nail them one by one – small grass-lined hollows almost indistinguishable from the grasses they lie among, each scraping marked with a tiny white flag to allow the wardens to locate them easily. Lapwings are remarkably beautiful birds. Their plumage looks a basic black-and-white in flight, the image I have retained from childhood at The Leigh, clouds of lapwings flickering above the Big Meadow. But now I have the time to study them on the ground through high-powered lenses, I can see that each bird wears a shining gorget around its neck and a coat of prismatic green embellished with thin scallops of white and a gleaming bronze shoulder-flash. The tall spiky crests at the back of their heads tremble in the wind. Their chicks run about like tiny toys, scudding into and out of sight, then crouching as motionlessly as they can.

Now I appreciate the need for the electric fence round the nesting area. A fox would play havoc with the lapwing colony if it got into Sutton's Marsh at nesting time. And although lapwings aren't exactly rare, their UK population is declining thanks to the nationwide loss of exactly the sort of damp, boggy, rough grassland habitat that Marshside provides.

On a fleet of water sails a male shoveler, a handsome drake with iridescent dark green head and chestnut flanks, a hefty blade of a beak and an eye like a drop of liquid gold. It is a peacocking get-up, designed by a cruel committee specifically to show up the drabness of his mate. She paddles nearby, subfusc to the point of self-effacement with her spotted brown body. The golfers on neighbouring Hesketh Links are far more gorgeous, displaying to one another in their turquoise shirts and kingfisher-blue trousers. I meet one of these 'other beings' in the hotel lift in Southport that night. His deep mahogany tan and wrinkle-defiant skin never came from a winter sun on these shores. 'Just spent the winter in Almeria,' he drawls. He flashes a china-white Hollywood smile. His *en brosse* hair is thick, raven black and parched-looking. He smells exotic in his double-breasted blazer and pressed slacks. He could be a carefully pruned sixty-five, or an absolutely shagged-out forty-five. He looks a downy bird, whatever.

Sixty miles or so north of Marshside, the first of the Lake District fells begins to rise. Here at Easter 1965 my father and I had our first holiday on foot together in the Mecca of English hill-walking. He hoped that I would somehow catch the fell-walking bug like measles and be infected with it for the rest of my days. It didn't put me off walking for evermore, but it was the nearest run thing you ever saw in your life.

Taking me away for a week *à deux* was an extremely unselfish thing for him to do, I see now, because it would be fair to say that Dad and I didn't see eye to eye at this stage of our lives. He was a year into his post as J, head of the Soviet Bloc section at GCHQ, up to his elbows in the Cold War at the moment when Lyndon Johnson was committing US combat troops to Vietnam for the first time. Dad was tense and enclosed, and had little slack to cut for a touchy teenage boy. I was unhappy at school, resentful at home, short of funds, bored of living out in the wilderness at The Leigh. I was fifteen years old and jiggy with it – or I would very much like to have been. It was the sixties, man! Everyone else was swinging, everyone else was smokin' pot and having a ball. I'd been to Lord John's clothing emporium in Carnaby Street and bought a belt at least six inches wide with an outsize silver buckle (it sat in a drawer from the moment I got it home, being far too big to go through any of the belt loops on my trousers). Roo was just sixteen, and had a motorbike on which we careered around the lanes. I'd been smoking like a chimney for a couple of years, starting with untipped Senior Service which I pinched two at a time (one for me, one for Roo) from the cigarette box kept full for guests by my non-smoking parents. I'd recently learned to swear, too, proper naughty ones. But I lacked the confidence to confront Dad with my full range of emotional turmoil. He was just too grown-up. He could demolish any ill-founded outburst of mine with an incisive statement of the facts.

'I'm going to get a motorbike.'

'Motor-bikes cost money.'

Doh.

My father appeared the squarest of squares with his short-back-and-sides haircut, his pressed trousers and polished shoes, his moral compass seemingly pointed unswervingly to 'sensible' and 'good'. I took for granted then, and only later came to appreciate as rather extraordinary, what most of my friends found it impossible to credit any parent with, Dad's really anarchic sense of the ridiculous. I suspect this was a way of syphoning off some of his anxiety, taking advantage of situations so crazy he couldn't possibly be expected to control them, and could therefore sit back and have a bloody good laugh at them. At such moments we children felt freed from the normal constraints and could join in without holding back. Hysterical laughing jags at family meal times, triggered by the dog farting or my little sister Lou giving herself a mashed potato hairdo, would have us literally rolling on the floor and laughing till it hurt. In the 1950s Dad had indulged this side of himself by buying comic records such as Peter Ustinov's 'Mock Mozart' and 'Phoney Folklore'. As children on car journeys my sisters and I would chant Ustinov's off-the-wall introductions to the songs – 'First, from Russia, ze song of ze peasant whose tractor has betrayed heem . . . Second, from Norvay, de song of de g-nome slighted by a dilatory troll, tee-hee.' But by the time I was a teenager the whole thing seemed leaden and unfunny. Had I known that the man who recorded 'Mock Mozart' was George Martin, now busy producing The Beatles and all their wonderful works, I'd probably have shrugged. So what? Could he play a freaky solo like Dave Davies of The Kinks?

Dad was not at all sympathetic to the music of the 1960s counter-culture. But his musical tastes, so apparently conventional and centred around Bach, Handel and Beethoven, had their more relaxed aspect, too. At the back of the cupboard in the study – another supererogatory room, like the dressing room, that served men of his background – I had recently discovered a cache of LPs sung by chaps with guitars. Singers and their guitars bulked large in my life just then. I had bought a cheap guitar and was learning to wrestle a few chords out of it, while over the airwaves came voices and tunes full of inchoate rage and restlessness. Admittedly Cliff 'Bloody' Richard sat atop the charts at Easter 1965 with 'The Minute You're Gone', but The Who had come jerking and windmilling in at

No. 8 with 'I Can't Explain', while Dylan ran hot on their heels with 'The Times They Are A-Changin''.

Fuck, yeah, as I would have said (well out of earshot) had I known the phrase.

The LPs in Dad's study cupboard, however, proved to be on obscure American labels, and to contain dust-bowl ballads and hobo songs such as 'Mule Skinner Blues', 'Beans, Bacon, And Gravy' and 'Hallelujah, I'm A Bum' by men with names like Cisco Houston, Woody Guthrie and Haywire Mac. I put a couple on the turntable when Dad wasn't around, hoping for a new source of inspiration. But the guitar playing seemed crude, the singing tuneless and the words irrelevant to my situation. Dad must have brought these droplets of American folklore back from stateside trips on office business. It pleases me now to think of him going round the record stores of Washington, DC. 'Have you got anything by Haywire Mac, please?' But how did he develop the taste for such arcana in the first place? Not for the first time in this year of edging closer to the man my father really was, I feel a powerful impulse to ask him a question, to hear his voice once more.

The black Morris 8, GYC 898, had long since bitten the dust by the time we went walking in the Lake District. Now we had a powder-blue Mini, 5756 AD. The motorway network wasn't yet complete; there was a great gap around Birmingham. It took Dad all damn day to drive from The Leigh to the Lakes, cramped up in this bite-size car beside a hulking teenager in the terminal stages of rebellious angst. We trundled north behind eructating lorries for hour after hour at an average speed of 30 mph, up the A38 to Birmingham, through the city centre and on along the A34 to Stafford and two or three hours of the newly opened M6. At the far end of that there was still an interminable farrago of hilly roads and narrow bends before we reached Borrowdale and the Royal Oak at Rosthwaite.

There, after we'd unpacked, Dad produced a large cardboard box. When he first went fell-walking as a young man, he said, he'd worn a pair of nailed boots – heavy as hell, and slippery on wet rock. But there were better boots now, especially for beginners. I'd never had proper walking boots, so here was my first pair.

My first boots! The box they came in was stamped diagonally in black Gothic letters with the single word 'Veldtschoen'. The name tasted both romantic and macho when rolled around the palate. Field shoes, eh? Upended from their box, the boots fell to the floor with a satisfyingly manly thud. I laced a ridiculous number of eye-holes, stood up with what seemed like a ton of lead on the end of each leg, and tried to step forward. The Veldtschoen were so heavy they remained anchored to the spot. I did not. Damage: second-degree contusions to knees and knuckles; massive injuries to teen self-esteem.

What neither Dad nor I knew was that Veldtschoen required at least a month's hard labour with kneading stick and urine spray (yes, really) before you put them on. I don't know whether the ox who died to construct them had lived a particularly hard life on trek, but the Veldtschoen were as unyielding as iron, and so weighty that Tony Soprano's mob could have saved him a fortune in cement by pinching a lorry-load. Those bloody boots were literally unwearable. Literally bloody, too – our initial expedition, a gentle four-mile stroll to Stockley Bridge and back on that first evening, gouged wounds in the backs of my heels so deep that I still carry the scars, fifty years on.

My feet stung and ached, but I still remember the sense of deliverance with which, on my own at last, I furtively opened my bedroom window in the Royal Oak late that night and puffed the first illicit lungful of Embassy into the rainy Cumberland night. Alone in his own spartan room next door, Dad's relief at being set free from incarceration with the sulkily silent glower of my teenage countenance and the smell of my teenage feet must have been even sweeter.

If there's a patron saint of fell-walkers, it's Alfred Wainwright. Can there ever have been a more influential series of walking guides than the seven hand-drawn, modest-looking little volumes of Wainwright's *A Pictorial Guide to the Lakeland Fells*? He published Book One, *The Eastern Fells*, in 1955, and six of the seven guides had appeared by the time that Dad and I arrived in Borrowdale.

Dad had no truck with Wainwright back then. It wasn't until ten years later, when he and I set off to walk the Pennine Way, that AW properly came to my father's notice in the shape of his *Pennine Way Companion*. From his twenties to his fifties, Dad had two Lake District walking bibles only – *The Thorough Guide to the English Lake District* by M. J. B. Baddeley, B.A. (1880), and *Walking in the Lake*

District by the Revd H. H. Symonds (1933). Baddeley's chunky little red book must originally have been bought by my great-grandfather. It dealt with geology, botany, train timetables, steamboats, breakfast charges and carriage drives, and contained a great number of recommended walks, long outdated and accompanied by all-but-useless maps. Symonds was better fun than Baddeley, far bossier and more didactic. He loved to tell you what to think, and on which crag to think it, often taking on the persona of AUCTOR for a caustic exchange with LECTOR about charabancs (unspeakable), girls in shorts and zebra stockings (utterly beyond the pale), and Oxford softies who had ruined the manliness of the current crop by insisting on the introduction of hot baths at Seatoller House.

The phraseology employed by Symonds was hearty and tasty, a beefsteak-and-ale way of talking:

'I counsel you to walk west with a spice of north in it.'

'Seen between Pillar and Scoatfell over the bowl of a pipe, Windy Gap is a eupeptic foreground to any man's lunch.'

Of a hotel in Wasdale: 'A gaunt habitation where rock-climbers frequent and talk shop grossly, and common folk wait in bed till their clothes dry. Here Owen Glyn Jones would do the traverse of the billiard table by the grip of his mere fingers along the cushion – so they say.'

Symonds was a muscular Christian, Anglican priest, teacher and pioneer conservationist, a founder of the youth hostel movement, secretary and chairman of the Friends of the Lake District, President of the Ramblers'

Association, secretary of the National Parks Committee of CPRE, and a supporter and catalyst for the establishment of the national parks after the Second World War. Undoubtedly a great man, a doer. And an incorrigible romantic and nostalgist, too, viz. this splendid passage from the chapter in *Walking in the Lake District* which deals with Borrowdale – 'Borro'd'l', as he'd have insisted on pronouncing it:

> Give me thirty-six hours of Borrowdale rain, when the clouds follow one another down on to the white houses of Seatoller, thrusting upon each other in black succession from the V-shaped opening of the 'Stee', and you may have your thirty-six hours on some rainless East Anglian beach, and I shall not envy you.
>
> For do you know Seatoller? and Seatoller House? and did you know Moses Pepper? – stern Rechabite, quarryman, and, by the loyal aid of wife and daughters, master of the most famous place of lodging in all Cumberland – and the family maintains it still. The visitors' book of that house, more than of any dalehead house, is a piece of history. Cabinet Ministers, undergraduates, professors, publicists, walkers, lawyers, runners – you will find them in that book. All King's and Trinity used to sleep there – all of them that could walk: did you never hear of the 'Lake Hunt' of past days? – when there were no motor coaches, and the day's letters jogged up in a pony cart, something

after twelve, and there were no hot-baths there
(Oxford's entreaties led to those), nor cold baths
either, but the same Elysian pool under the big
rock, where there is a foot-bridge to-day and
curious eyes? In those great days of old, men *ran*
upon the fells, chasing paper and fleet long-
distance champions and high hopes of dinner in
the evening. And Seatoller is a grand place still to
be in – but for all the high hours of a fine day now
you must be away, well up above reversing motor
coaches. In the evening the sunset lights scatter
high colour, as before, on High Knotts and the
ridge of Chapel Fell, and the snails walk abroad
stoutly in the garden.

Splendid stuff for right-thinking chaps.

H. H. Symonds deeply disapproved of cars. For him there
was only one permissible way to enter the Lake District –
on foot, walking in over several days from the railway
station at Shap. His shade frowns on me today as I park
my car in Elterwater, but gives my early start (8 a.m.) the
ghost of an approving nod. I aim to walk over Lingmoor
Fell, a hike I've never done on a fell that few walkers
bother with. Symonds barely mentions it. But Wainwright
allows Lingmoor a few pages in Book 4 of *A Pictorial Guide
to the Lakeland Fells,* and declares that the summit is a fine
vantage point from which 'the surround of rugged heights
towering above the valley head of Great Langdale is most

impressive'. That's good enough for me. I'll follow the Master's footsteps across one of his meticulously drawn maps.

Walkers are already trickling into the village under a blue sky streaked with white; tough youngsters toting big packs and serious boots. I find a stone-rubbled cart track climbing away south-west between woods of hazel, larch and oak filled with wood anemones. The mossy walls are footed in banks of violets and opposite-leaved golden saxifrage. This far north, the bracken tips are still tightly curled into fiddlehead scrolls. A big old pine droops its tattered branches half open like an oiled cormorant.

The path rises over blue-green slaty rock thick with barren strawberry and celandines. There are shoots of foxglove leaves not more than half an inch long. The sun encourages a faint herbal smell out of the crinkled leaves of wood sage. Orange tip and peacock butterflies bask on warm stones. The first lizard of the year flicks away from the damp mossy seep where it's been drinking, and is out of sight among the rocks in less than half a second. Above Dale Head Farm I turn off up a fellside path, stony and clear, snaking uphill under crags and through old quarries and tumbled stone walls. Little Langdale Tarn opens up below, as dark as ink. The trees on the slopes of Moss Rigg beyond are still naked, their skeletal profiles unthickened by any spring growth. They look to be at the same stage as those I saw two months ago above Rodney Stoke in Somerset. Back east the crumpled slopes around Wansfell Pike rise on the far side of Windermere, foregrounded by Elterwater and wooded Nab Island.

At the top of the climb a stone wall leads boldly west – a wall in excellent repair for once, with coping stones sloped at 45 degrees, a beautifully maintained work of art as much as of practicality, advancing up seemingly impossible slopes and running precipitously down into gullies like a lakeland version of Hadrian's Wall. The path beside it has been trodden to a fine crispbread of pale grass, flattened by the winter's snow. The plateau it crosses is still waiting for spring, still devoid of flowers. I stop to savour the silence, and hear the hum of a bee, the tiny whine of a hoverfly, the faint snick-snick of a Herdwick ewe's teeth as she snips and tugs at selected patches of grass. A sharp *chizzit!* calls my attention to a rock where a meadow pipit stands preening its breast, the beak dividing the pale spotted feathers to reveal a dark, silky layer of down underneath. Overhead comes the rattle and rush of raven wings, and I look up to see the big black bird putting on a sky display, gliding with wingtip fingers widely separated at maximum stretch, then suddenly pulling back his wings into a vee as he ducks his head and swoops sideways. A sideslip, a barrel roll; then he resumes his glide with a two-tone chuckle: *Haw-haw, huh-huh*. There must be a female watching him with admiration from a nest nearby.

North across the Langdale valley rises the jagged profile of the Langdale Pikes. These grey and brown fell tops sit in line across the top of lumpy Langdale Fell like gnarled old men on a settle – humpbacked Pike o' Stickle the westernmost, its neighbours the spiky-headed Loft Crag, Thorn Crag crouched next under the bare upraised molar of Harrison Stickle, with the tortoise neck of Pavey Ark poking off eastward at the end of the line. As I watch, a paraglider jumps from the crest of Harrison Stickle and goes swooping along the line of cliffs that buttress the Pikes, an impudent simulation of the raven's air mastery. I think once more of Giorgos Papoutsakis, son of my friend Lambros, caught in an updraft off a Cretan crag five years ago, Icarus in the moment before the fall which killed him then, and his mother Maria by slow degrees thereafter. And I make a promise to Maria and Giorgos that before the year is out I will go to Thronos to drink and sing with Lambros beside his raki still.

Lingmoor Fell is neatly divided by a wall, with pale grass to the south and dark heather moor to the north. I walk up to the cairn on Brown How, but it's already occupied by a pair of walkers. I feel an ignoble stab of resentment. Get off! I wanted to stand there!

I descend along the stone wall towards Langdale, where later that afternoon I'll catch the bus back to Elterwater among a crowd of pungent peak-baggers, steaming fresh from their triumphs on the Pikes. A side track catches my eye on the way down, and I accept its invitation to turn aside to Lingmoor Tarn. The path is no more than a faint sheep trod. Halfway to the tarn I see in the heather a pair of black-rimmed eyes an inch apart, staring at me unblinkingly from under a soft white fringe of fur. The hair rises on the back of my neck. Then my own eyes come into focus, and I see that it's the wings of an emperor moth – not the technicoloured 'bull's face' pattern of the male, but a female moth's subtler tones of ivory with ash-grey scallop lines, a marking on each wing like a mournfully drooping eye with tawny iris and velvet black pupil, and a tiny crimson streak like a flame at either wingtip.

Lingmoor Tarn with its minuscule islets lies beyond, cold and still. The water is pimpled with reed stems. Three slim and stunted silver birches grow out of the central islet, a grouping as natural and yet graceful as though it has been arranged for some Japanese landscape painter. The tarn lies among hillocks in a sheltered hollow with no one in sight. A bit later in the year I'd go in for a skinny dip; there's nothing so good as the silky feel of peat-infused water sliding against the skin. But the water of Lingmoor Tarn looks too shallow and cold on this April day.

Next evening, descending from Kentmere Pike in the eastern skirts of the Lake District, I pass a rookery at Hallow Bank, ten nests built very close together in the tops of two adjacent trees. As I walk underneath, a stranger rook comes flying in to land on a branch in the heart of the community. All the inmates immediately explode into admonitory cawing: *Aaaah! Aaaah!* One adult bounces off its nest and squares aggressively up to the intruder with outstretched neck and wide-open beak: *Aar, aar, aar!* And the stranger stands its ground, head forward, beak agape with defiance: *Aaagh! Aaagh!* Three distinct timbres of call, each with its own purpose.

In the lane beyond, an old man is leaning on the wall, inspecting his sheep in the evening sunlight. With flat cap, stick and thick glasses he looks every inch the retired farmer. 'Texels and Cheviot crosses,' he says. 'Just looking them over. We've had a good lambing this year. Worst thing for lambing is when it snows. I remember one Friday it was snowing when we got up. We thought it'd be over by ten, but it kept on. Saturday morning, it was still snowing. Sunday morning, still snowing. The chaps came in and said, "We can't see sheep nor lambs out there." I said, "Go and find what lambs you can and we'll bring 'em in." We had forty inside, jammed up tight together, bottle-feeding 'em. Monday morning, snow had cleared – all gone. So we'd forty lambs in the shed, all smelling the same as each other, to match with forty ewes on the hill that had lost them! We put neatsfoot oil on all the lambs and all the ewes, so none of 'em could tell t'other from which. It were a job – but we got them all paired off. Got shot of the lot of them!'

Back at the car, pulling off my boots, I notice that one of the metal eyelets through which the bootlace passes has pulled free from its hole. The boots are less than a year old, and were made in China.

My week with Dad in Borrowdale didn't turn out too badly. Patched up with Elastoplast and wearing a new pair of hastily bought cheapo boots, I winced up and down Red Pike, Haystacks, Catbells and Scafell Pike. The Four Passes walk proved a step too far, though I was quite proud afterwards of the blood on my socks. We got lost time and again trying to follow M. J. B. Baddeley's long-out-of-date advice – lost and soaked. Somehow I wasn't put off for life. The sneaky pack of Embassy helped. So did the arrival at the Royal Oak, on our second day, of a girl just a bit older and even sulkier – but in a really good way.

Dad was no hearty jock, forcing me out in the open air to do me good. He'd hated his own teenage years at Dartmouth naval college in the 1930s – the bullying and teasing, the anti-intellectual atmosphere, the fact that the son of Rear Admiral James Somerville couldn't climb a rope or tackle a fly-half or catch a cricket ball. But walking in the Lakes was one of the portals through which a boy needed to pass in order to become a man, like learning to use a shotgun or paint a bedside cabinet, make a balsa-wood glider or tie a reef knot (right over left, left over right). The love inside men like Dad was practical in expression, thoughtful and modest. Love for a son was not quenched by teenage rejection, though it was sorely tried. Love was to be demonstrated in deeds, not words. The nearest he got to an endearment was the 'Dear old boy' with which he always began his letters to me. He'd never have thought of saying, 'I love you, son.' And I wouldn't have wanted, then, to hear it from a man like him.

May

The man of May stands very still, watching
the children dance away the day . . .

W E SET OFF FROM our house in Bristol at 3.15 a.m. in complete darkness. The weather forecast suggests a clear sky, with perhaps a bit of cloud in the east – exactly what we have been hoping for. We're encased in thermals and have a flask of tea-plus-something with us, as well as a bag of gloves, scarves, hats and walking boots. It's 4°C, cold enough to snow if this were 1 March and not the First of May.

The motorway drive goes quickly. Away in the west beyond Gloucester the sky is black, and there's no sign of the symmetrical whaleback of a hill that I named 'the Beetle' when I was four years old. It was a private word, but I knew what I meant. A semi-ovoid clump of trees stood at the crest of the Beetle. In my childish eyes the spinney was a woodlouse, frozen in the act of crawling across the hill. Woodlice were objects of fascination. They could crop up anywhere, but the one place Roo and I could be sure of unearthing one was inside the half-rotted planks that formed the walls of the coalshed at Hoefield House. Prise up a splinter and there they would be, shiny black or brown, sometimes curled up tightly into a ball, sometimes hurrying away back into the dark with that distinctive clockwork-toy motion they had. They couldn't hurt you, we found out. A woodlouse couldn't bite or sting. It tickled, though, as it bumbled across your palm or up your arm, multiple legs rippling under the skirts of its jointed carapace, whiskers questing right and left. Woodlice didn't seem to have eyes or ears, mouths or even faces, but they had character and purpose. They didn't hesitate or panic. Either they clenched themselves into a tiny, impenetrable fist of plated armour, or they busily got themselves out of the light and in where we couldn't follow.

When I became a man, I put away childish things – some of them. Now the Beetle is May Hill, a familiar lump on the western horizon whenever I travel north on the M5, and the woodlouse of a spinney on top has gained the name of 'The Ploughman and His Team'. Many of the trees are Corsican pines planted in 1887 to celebrate Queen Victoria's Golden Jubilee, and others are their younger distant cousins, Scots pines planted in 1977 for our own Queen's Silver Jubilee. Among the youngsters stands one gnarled and lightning-blasted old Scot who must antedate the Golden Jubilee. Drovers loved to plant tree clumps to mark the site of good grazing, preferring evergreens for their height, darkness and distinctive forms that were easy to spot in the landscape. In fact, the hill seems to have had a clump of trees since the Civil War, and maybe long before that. But not for much longer, perhaps. The Corsicans are suffering from Red Band Blight, a fungus which defoliates the trees. Increased spring and summer rainfall and warmth are to blame, it's said.

The Ploughman and His Team sounds like a name drawn from the deep wells of time. But in fact it was John Masefield, writing in 1911 of the southward view from his native Ledbury in *The Everlasting Mercy*, who dreamed up

the notion of the clump resembling a ploughman following a team of oxen.

Near Bullen Bank, on Gloucester Road,
Thy everlasting mercy showed
The ploughman patient on the hill
Forever there, forever still,
Ploughing the hill with steady yoke
Of pine-trees lightning-struck and broke.
I've marked the May Hill ploughman stay
There on his hill, day after day
Driving his team against the sky,
While men and women live and die.
And now and then he seems to stoop
To clear the coulter with the scoop,
Or touch an ox to haw or gee
While Severn stream goes out to sea.
The sea with all her ships and sails,
And that great smoky port in Wales,
And Gloucester tower bright i' the sun,
All know that patient wandering one.
And sometimes when they burn the leaves
The bonfires' smoking trails and heaves,
And girt red flamës twink and twire
As though he ploughed the hill afire.
And in men's hearts in many lands
A spiritual ploughman stands
Forever waiting, waiting now,
The heart's 'Put in, man, zook the plough.'

*

The Ploughman and His Team . . . As a local boy I never
heard that romantic name with its stellar and ancient res-
onances. And we never took part in this ritual expedition
I'm embarked on today, one I've been promising myself
I'll join, year by faithless year – the climbing of May Hill
on May Day to welcome in the dawn. I never knew Dad or
Mum to do it in all the time we lived at The Leigh. Maybe
it wasn't even done in their day. Stories say it *used* to
be done, more than three centuries ago when May Hill
was called Yartleton Hill. Back then at sunrise on May
morning the youth of Newent would stage a mock battle
between winter and summer on Yartleton summit, with
summer always victorious. Then lads and maidens would
carry spring greenery and flowers back to their houses.
What a lovely and innocent picture. No mention there
of belly-bumping, boozing or bullyragging, activities
common to hilltop rituals at the spring of the year since
Beltane was celebrated across the pagan land of Britain.

There's still only a suspicion of lightening in the east at a
quarter past four when we reach the scattered village of
May Hill. Or is this trail of houses spaced out round a
loop of roads truly called Gander Green? The road signs
seem unsure. Where is the village hall, folk want to know.
They follow other cars along the dark lanes, stop and
wind down the windows when they spot anyone on foot.
D'you know where we're supposed to start from? Nervous
giggles. It all looks the same round here, eh? We need a
local to materialize and point out the way. But the villagers are tucked up in their beds, snug and smug behind

their curtains. The houses with their blank averted eyes look as though they know something. But out here in the cold night we're all strangers.

The three of us park the car where we guess the main track up May Hill crosses one of the lanes. We click the doors shut as gently as possible and set out to climb the hill by torchlight, treading as quietly as we can on the clinking flints of the track among the trees. There's the sound of muted conversation somewhere behind us, but that soon fades away. Every bird in these woods is silent. There's only the sound of our breathing, the faint creak of boot leather and the glassy tinkle of the stones. Then ahead a dog barks, and a blackbird breaks out scolding. It turns to tentative notes, sweet and unsure. A wren whirrs briefly. A robin begins to chitter, and deeper in the wood a warbler produces some sweet, expressive phrases. By the time we leave the edge of the wood and enter the common land of May Hill top, the dawn chorus has got under way. There's another musical sound, too, faint but growing louder, coming up behind us – the silver jingle of tiny bells, bound round the shins of three men who are walking the hill in ribbon coats and breeches. They prove dour but cheery, as only folkies and morris dancers can be. These are the Forest of Dean Morris Men, come to dance the sun up. One carries a Green Man stitched to his coat, peeping out from a thicket of ribbons. Their jocularity and gruff byplay can't disguise the slight whiff of the sinister which always attends such dancing and mumming troupes with their rags, tatters and lumpy-bumpy music.

Around the summit of the hill there's a trace of an ancient ditch, which encloses a tump that might be a round barrow and some shallow pits now sunk in the grass. There are rumours of tunnels, tales of buried treasure. But May Hill seems never to have been formally excavated. Now, though, at quarter to five on this freezing cold morning, there's a faint seashore roar of wind and a tang of woodsmoke among the trees as the Ploughman and His Team bring forth shadowy figures into the first light of May morning – a woman swathed head to toe in a Victorian heroine's cloak, stumbling yawners, some kind of mage with tangled white locks who stands at the edge of the trees looking nobly east. The clump is crawling with overnight campers and their tents.

We stand and talk to a man of my age with a full white beard, who wears leggings and highly polished shoes under a wonderful old work smock. It's a beautiful piece of craftsmanship with a broad collar and cuffed sleeves, the smocked front edged with star motifs. 'It's the genuine thing,' he says, 'probably a hundred years old or more, I shouldn't wonder. I got it from a thatcher in Essex.'

A thin trickle of music from a single fife calls our attention. A bunch of morris men, rubbing their red hands and slapping their arms round themselves, are getting ready for action, silhouetted against the gradually broadening light in the east. 'We are the Forest of Dean Morris Men,' cries the foreman. 'All the way from . . . here!' We titter and blow on our frozen fingers. 'That was a traditional calling-on tune,' the foreman shouts in ripe Forest accents, 'collected by Cecil Sharp in Clifford's Mesne a hundred years ago.' He doesn't divulge its name, which is 'Cuckolds All In A Row'. The hum of a melodeon follows, and the Forest of Dean get going. They shuffle and kick, they leap and hey and ho. It's galumphing music, but curiously graceful patterns emerge as you watch and listen. The Lassington Oak in their leather weskits follow on – or maybe they do; it's hard to distinguish between the troupes of middle-aged men. A women's side sprints in, their sturdy calves encased in stripy stockings, and they bring new vigour to the dancing as they gallop in and out. Upraised handkerchiefs flutter against a sky that has suddenly filled with dawn light. The clouds low down on the eastern skyline flame up into a smear of rose and gold. The first crescent of the May Day sun slips out from below the rim of the hills and illuminates a surf of mist over the pastures and orchards in the valleys below.

The Three Counties (Gloucestershire, Herefordshire, Worcestershire) are famous not so much for beer as for cider. They used to be celebrated for perry, too, made from the bitter semi-wild pears that are cultivated, these days, in decreasing numbers. In the Three Counties, old stories said that all the pear orchards originated from one tree that grew on May Hill. One day a hungry fairy man helped himself from the tree, but the fruit proved too bitter. Revolted, he spat out his mouthful, and all the seeds it contained, right across the Three Counties. Cue much perry-making, and merrymaking.

These days you don't get much call for perry

hereabouts, or anywhere else. Perry is a drink that occasionally defines fashion, but usually defies it. As children we used sometimes to drink a very sweet and fizzy commercial version, a treat on high days and holidays, utterly different in every way from the beautiful mature perry called *most* that I once sampled in an Austrian cheese hut. That Bregenzerwald version of perry had been sitting on the lees in its barrel in the cellar beneath the hut all winter and through the following spring and summer, gently cooling and gently warming with the seasons, slowly gaining flavour, strength and subtlety. It was deliciously soft and cool in the mouth. But it was the image of the wooden barrel and its pale gold cargo among the wheels of cheeses, all unhurriedly ripening under the earth, which infused the *most* with magic.

Though the dancing on the hilltop hasn't finished, we have stood still and frozen long enough. God knows that five-thirty in the morning is no time to be drinking beer, but over there at the edge of the clump there's a fairy man with a table full of the stuff. 'Go on, it's free,' he urges us. 'Hillside Brewery, just got started. See what you think.' We think it's just fine. A bottle of Legend of Hillside (highly appropriate) and another of Legless Cow (ditto) in hand, and we walk away from the dancers and the ring of onlookers, heading north along the boggy edge of May Hill among bluebells and violets and frost-rimed grasses. Down through Bearfoot Wood, where the saw-edged leaves of sweet chestnut are just emerging on trees last coppiced twenty years ago. We tilt the beer bottles to our mouths and walk among heads of wild daffodils already withered and dry, and the plump white bells of wood sorrel, with the dawn chorus reeling out all round.

At the foot of the hill where we have danced and drunk the May in, the houses are silent, the bedroom curtains still drawn. In the village hall we find trestle tables strewn with apple blossom, a bustle and a great smell of breakfast, and the Lassington Oak and the Forest of Dean and their Green Men queuing for fry-ups. It's hungry work, dancing up the sun.

Notes from the Old Bristol Road, 10–15 May 2015:

Cloudy, 13°C.

Trees all in leaf, buds broken – except for ash leaves, only halfway out. Chestnuts in flower. Hazel seeds set. May/apple blossom fully out.

Cleavers have spread right up the hedges. Big dock leaves. Grasses still sleek – haven't coarsened and spread yet. Plantains have black seed heads now.

White flowers – wild garlic, stitchwort, Jack-by-the-hedge (already bolting), cow parsley, white dead-nettle, ox-eye daisies.

Red/pink flowers – campion, herb Robert; red valerian growing out of wall cracks.

Blue flowers – bluebells, forget-me-nots, bugle.

Yellow flowers – hawkbit; buttercups, daisies, cowslips by Chew Valley lake. Dandelions mostly now bald or going into clocks; stems starting to turn shiny pink.

Lambs now half the size of their mothers. Half-grown young deer seen in field.

On Priddy Pool's gruffy ground (disturbed by lead mining) – tufted grasses beginning to green up, but no seed heads yet.

At the foot of the wall opposite my mother's house, a hen pheasant waits. Beside her crouches one tiny chick. Another chick is perched on top of the wall about man-height above the pavement, too scared to descend. The hen flies up beside the laggard, then flutters back down to the road and gives a soft screech of a call, as though to say, 'That's how it's done – come on, now.' There is a flurry of feathers, and the squatter throws itself off the wall and comes tumbling down into the road, rolling over and over before righting itself and scuttling off, apparently none the worse. The equivalent fall for a human would be about a hundred and forty feet.

In Mum's garden walks another hen pheasant, as drab as can be, bundled up and bulky in her puffed-out brown feathers. Displaying to her, a male in all his shining glory – burnished bronze back glinting in the sun, ginger belly, bright scarlet face and erect ear tufts. He stalks her, strutting at a slow march to the front of her, cutting off her retreat. Then he puffs himself out, lowering his face almost to the grass as though in self-abasement, and fans out his tail vertically, shivering it like a Japanese gentleman's fan, each feather widely separated from its neighbour. The hen suffers him to get a little closer, then gives a mighty hop and scuttle that lands her several feet away. The cock is left to repeat the whole process, over and over again. At last he loses courage and purpose, hunches his shoulders and limps away with the bowed back and plodding gait of an exhausted, put-upon suitor.

Five minutes' recuperation in the long grass and he is out on the lawn, picking up seeds as though hanky-panky were the last thing on his mind.

A morning of spitting rain and cold wind, three hundred miles north. The Durham dales roll enticingly, bisected by the River Tees. I'm walking upriver for a few days, that's the plan, heading north-west to where the Tees topples over a series of waterfalls among rock ledges jewelled with an Ice Age flora. What strikes me just now is how Barnard Castle has changed from the grey, run-down little town I used to know when I was a student in Durham. 'Barney' was a place to walk warily on a Saturday night, a local place for local people, where lads hanging out at the market cross would bray you if you gawked at them wrang. Something has happened to Barney since I was last here – something called 'the middle classes'. The town is bustling. People still shout at you in the street, but it's only 'Hello!' Goat curry pasties and artisan bread are on sale at The Moody Baker. As I pass the shopfront of William Peat, Master Butcher, a man in an apron comes dashing out after a customer, waving a greaseproof packet: 'You forgot your sausages, missus!'

The castle is all tall curtain walls and drum towers, a big hollow box of stone hanging over the Tees. Down by the river, the rustling bar of the weir flashes with light. A dozen martins are curving and swooping here over a sunlit patch of water, hunting flying insects beyond the wall that stiffens the bank. A pair is in chase, about five lengths apart, the hind bird mirroring every twist and turn of the leader in manoeuvres reminiscent of grainy film clips of Spitfires chasing Messerschmitt 109s.

As I stand and watch, I see out of the corner of my eye a martin fly at top speed straight into the wall below me. A closer look shows the mouth of a pipe about six feet above the water level. The martins must be nesting in there. What they'll do if there should be a cloudburst and one of the springs in the steep bank above needs to relieve itself through the pipe, I can't imagine, but perhaps they know that the pipe is blocked or disconnected. Every minute or so, one or other of the parent birds detaches itself from the whirl of martins over the river, makes for the wall and flies headlong into the pipe with a beakful of insects. Five seconds later the same adult, having thrust its cargo down a nestling's throat, emerges with a tiny white dropping in its beak. Half a second to put some distance between itself and the nest, and the dropping is released to fall like a bomb into the river and instantly dissolve in a milky cloud.

A flicker of white and a dipper has alighted on the concrete sill of the weir, bobbing and ducking its pristine shirt front before skimming off up the river in a flightpath as straight as an arrow.

Through steep-sided woods of young beech and lime the Tees comes down, silent and steel-blue in its deeper stretches, loud and toffee-brown over the shallows. The riverside path is a pure delight to walk. The range of spring flowers is remarkable, all across a rainbow spectrum of colour: red campion, herb Robert and blush-pink water avens; a pale milky fuzz of woodrush and white stars of woodruff, stitchwort, wild strawberry and garlic-scented ramsons; yellow flowers from the sherbet tones of primrose through celandine, gorse and dandelion to goldilocks buttercups; intense blue of forget-me-not and speedwell; and the pale mauves of wood cranesbill and wood anemone, all the way down to the rich hue of the violets on the mossy banks. These flowers are all in full bloom at the same time, taking advantage of a growing season that's shorter and cooler than down in the southern part of the country.

A lively sky stretches up ahead, white and silver altocumulus clouds piled like whipped cream on a blue plate. Black-headed gulls screech angrily from the riverside. The Tees divides noisily around a long islet of flat grey shillets, where a curlew goes with fastidious steps among the stones. Miniature sandstone cliffs outcrop along the valley slopes, and the beech trees, gripping them with roots like talons, are engaged in the slow but sure business of forcing the strata apart and crumbling the rocks to dust. Looking at the glacial pace of the process, the silent strength required, it's tempting to imagine a will at work among the beeches, an impetus to advance one inch at a time until they have gripped every cliff in Teesdale to death.

Unlike the beeches, I don't seem to be able to take my time today. I'm hustling and hastening as though late for a train. There's no need for it at all. I have a word with myself. Why are you marching along, fool? I make myself sit down on an ancient bench, rustically knobbled in cast iron, to spend five minutes contemplating the haze of blue and pink blooms under the trees. But two minutes later the itch in my soul drives me on, steeply up through the trees to stride through fields full of fat white sheep around East Holme Farm. A blunt-faced ram stands foursquare in the path. He doesn't budge at all as I hurry past, just puts his snub nose out and tests my wind, his golden eyes with their horizontal black streaks of pupils flicking ceaselessly between his ewes and me.

I drop down a steep primrose bank to cross the Tees and a rushing tributary stream. A fisherman stands waist-high in waders in the river, intent on the floating loops of his line. He doesn't look up as I go by. Two pied wagtails flirt their tails on the stones. I might as well not be there for all the attention they pay me. It is as though I have suddenly grown invisible, a ripple on the surface of the day for the wind to unmake and remake and smooth away once more. I walk on more calmly, past a railed enclosure beside the Tees. Abraham Hilton of Barnard Castle – tea, wine and spirit merchant, philanthropist and benefactor of the poor, died 1902 – willed his remains to be buried here. He was a man of strong religious belief, but he didn't hold with the division of faith into separate denominations. He deliberately chose to lie in this unconsecrated place, six feet of ground unclaimed by any sect. Looking up the river and round at the meadows, I can see exactly why Abraham Hilton thought of this spot. The river and the moment are all that matters. It's a moment I recognize, a rare intimation of how very lightly a human existence weighs on the world. Is this what I have been trying to evade with all my striding today?

I stop my yomping. I slip the gearstick into neutral and coast along. The sky turns leaden. A tumble of freezing rain hammers down, a howl of wind and a crash of thunder, a spatter of hail that crackles in the trees. Winter is not quite done and dusted this far north, it seems. And now I'm free to speed up once more, no longer fleeing my own mortality but pushing forward into the moment, the hail on my face and the wind in my ears, the peaty smell of the rain-lashed Tees, and the oak trunks flushed to gold against a slate-grey sky by a brilliant chip of sun coming up from the west.

Past Cotherstone the river path turns tricky, winding now high at the edge of slippery cliffs, now low beside the water among great blocks of fallen stone. I cross spillways and sluices above the loudly rushing Tees, which has gained force and volume unbelievably quickly with the downpour. The river is bottle-brown, suffused with peat particles washed from the moors above, and it gives off the exhilarating whiff of turbulent water supercharged with oxygen beaten into it by vigorous descents of rocky rapids. To walk such a river is a hypnotic thing, and one can easily understand the tales of travellers mesmerized into losing their footing and falling in. In a mile or two the path leaves the Tees and goes through the walled and hedged pastures around Low Garth Farm. When I last came this way ten years ago the old farmhouse was in sad disrepair, and things haven't got any better. Actually I'm rather glad to see that Low Garth hasn't been got at. I've always assumed that this fine traditional dales longhouse in its sublime position by the river would have been bought by a stranger and tarted up to the nines. But the house stands as silent as I saw it last, windows shuttered and chimneys cold, the byre roof in holes, the walled yard a bristling thicket of hazel whips.

Up in Romaldkirk the church of St Romald stands back from the village green. Romald himself was one of the grandsons of the mighty Penda, King of Mercia in the turbulent times of the seventh century when king strove with king for mastery of each other's lands. As lives go, Romald's was brief but eventful. No sooner had he sprung out of the womb than he declared himself in good Latin to be a Christian, and demanded to be baptized. The only receptacle suitable for use as a font was a distant quern-stone, too heavy for any man to carry to the hut where Romald lay. But the infant's suggestion that prayer might lighten it enough to be lifted proved accurate, and he was duly christened. On his second day of life the miraculous babe preached a sermon on the Holy Trinity and the virtues of the Christian life, and on the third he died, his work completed.

St Romald's church is known as the 'Cathedral of the

Dales' with good reason; it's a magnificent building, solid and yet light, like the baptismal stone of its saint. It boasts a devil's door, a tall thin doorway in the north wall that was blocked up with stones in medieval times to keep the Devil out of the sacred building. In the north transept I find Hugh FitzHenry, Lord of Bedale, Ravensworth and Cotherstone, recumbent on his chest tomb, armoured in tremendous detail. He died in 1305 of wounds sustained while fighting for King Edward Longshanks against Scottish foes whipped up to rebellion by William Wallace. The Lord of Bedale lies flat on his back, his mailed right arm reaching across his body in the act of drawing a long, and slightly bent, sword from its scabbard. His is an attitude full of defiance. I'll never know whether Hugh FitzHenry's face reflected this fierceness, because his expression has been worn into inscrutability by the inquisitive hands of seven centuries. I can't resist the urge to pat the gallant knight's armoured cheek in my turn, another tiny touch on the way to smoothing him entirely out of existence.

Outside the church a young black Labrador comes leaping up, full of beans and plastered in mud. He won't be smoothed out of existence any time soon. He jumps all over me, despite the apologetic curses of his master.

'Bonnie, get down! I'm sorry, he's— *Get . . . down*, blast you, Bonnie! – he's only six months old. Oh, Lord, you're— Bonnie, *stop* it, leave the man alone! – you're mud all over. Sorry!'

I get Bonnie out of my face, not unkindly, brush off some of the mud and check my watch. Five or six miles done, maybe four to go. Yes, time for a quick one, I think. The brightly polished brass sneck of the Rose & Crown's front door gives a pleasingly precise little click as I lift it. Come in, do, we've been waiting for you. The log fire and the wooden settle, the steak pie and the Black Sheep bitter are in an ongoing conspiracy to confound all walkers' plans, but today I'm going to resist them like a . . . well . . . like a broken reed, it turns out. It's late in the afternoon by the time I stumble out into the wind and rain, cheeks aglow. Check the watch. Can't believe it. Try to work up a head of steam along the trackbed path of the old Teesdale Railway, but keep yawning and dragging the feet. Fat sheep in the fields, green grass in the cuttings. And heavy eyes in a hot face. Effortful walking – painfully so. It takes an hour for the cold wind to clear the firelight and beer fumes out of my head, and by that time I'm practically in Middleton-in-Teesdale. Give myself a bit of a talking-to. No more lunchtime pies and pints, OK? Yes, OK. Probably.

Forty years ago, almost to the day, Dad and I were suffering another set of beer-related problems in Middleton-in-Teesdale. It was a summer night in 1975. Ten years had passed since I had limped up and down the fells of Borrowdale as a sulky and footsore teenager at my father's heels. Things had changed a lot since then. I'd been to university (mind-opening) and done a stint of Voluntary Service Overseas in Papua New Guinea (eye-opening). I was a married man now, with a teaching job and a baby. Dad was in his late fifties and coming up to retirement from GCHQ. He'd been promoted to the Directorate at the end of the 1960s, and now had two more years to complete as DO, Director of Organization & Establishments. This very senior post put him in charge of the divisions concerned with personnel, finance and security, and gave him a wide influence in the whole organization. He carried a heavy sack of responsibility and worry, and, as ever, it was hard walking that helped relieve his burden. I felt we were long overdue a rapprochement. It was high time to find a door we might open between us, and our mutual love of treading the hills seemed to offer a key. A long-distance walk in some tough country, a few days tackling a bit of a challenge together, and we might start talking more openly to each other. The Pennine Way long-distance path looked likely, especially the seventy miles between Middleton-in-Teesdale and the Northumbrian borderlands at Hadrian's Wall. Neither of us knew these northern hills, but we both loved the way that John Hillaby had written about them in his beguiling account of his Land's-End-to-John-o'Groats walk, *Journey Through Britain*.

Dad bought Alfred Wainwright's *Pennine Way Companion* and sniffed over the route. My job was sorting out B&B accommodation for the five days it would take us to do the walk. Why I turned to the *Good Beer Guide* for advice, I can't now recall. Dad wasn't a pub-goer. But he and I had grown unaccustomed to being companionably in each other's company, and perhaps I thought a few pints might oil the wheels of some potentially awkward evenings. Whatever the reason, my choice for our first night in Middleton could not have been more misguided. The King's Head was the dingiest of the town's dingy pubs. It crouched below road level, its matt-black paint peeling. The welcome was friendly enough ('Hello, lads. John, Chris, don't just stand there, come in, eh!'). But there was rubber beef and rubber Yorkshire pudding for dinner in the company of a loud, gigantic colour TV, and a bedroom that must have been recently added by a DIY enthusiast – not a very skilled one. It was nothing more than a hardboard cube tacked on to the room next door. Its lower edge stuck out across the hairpin angle where the stairs turned, leaving a waist-high clearance under which Dad and I had to crawl like miners in a two-foot seam. The lighting came through a tiny hatchway, cut out of the party wall near the ceiling. Its source was a 40-watt bulb in the adjacent room. In that room slept a kind and sociable but enormous couple from Birmingham whose snores entered our room at unmitigated volume, like a pair of ogres pretending to be motorbikes.

Dad said nothing, but I saw him flinching as the lady squeezed herself into her chair next to him at breakfast the following morning (rubber eggs, rubber bacon) and asked if he wouldn't care for a rub-down before his walk. We had a damn good laugh about that once we'd put the town behind us, and in fact the farcical aspect of our dodgy night at Middleton-in-Teesdale proved a great ice-breaker between us for that first day on our own. Forty years on, as I walk out of Middleton on a bright spring morning, I find myself still chuckling. Would I put up with such accommodation, nowadays, with as good a grace as my father? Hmmm.

The Pennine Way has declined in its attraction to walkers, especially young ones, since Dad and I followed its muddy and – back then – ill-marked course. It is not challenging, or pretty, or exciting enough for folk who have tasted the dramas of the Coast-to-Coast, or can go trekking in the Himalayas or Iceland at a cost in time and money that's not really much greater than they'd expend for three weeks' footslog through peat bogs and over the bleakest of moors in guaranteed mist and rain. That's exactly why I love this path, the oldest of Britain's national trails. In all its mire and muck, its great shoulders of featureless moorland, the loneliness of its uplands and harshness of its weather, it shows the backbone of England as it really is – rough, unadorned, a place for facing fears and overcoming them, a place for adventures, heartbreakingly beautiful in its simplicity of water, rock and hill, as bracingly real as a slap of hail and a bluster of wind. Alfred Wainwright in his plainspoken bluntness is the perfect guide for this no-nonsense trail. No stone wall, no side path, tumbledown shed, gate or cairn goes unmarked on his beautifully hand-drawn and illustrated maps. He is utterly, meticulously reliable. That's not to say that Wainwright is humourless. Quirks, quillets and quiddities are his stock-in-trade, dry asides his watermark. Naturally his *Pennine Way Companion* is to be read from back to front, each page from bottom to top. That's the way one reads a map, you see. And in case you need encouragement, laconic little aphorisms sum up what to expect from each stage of the walk:

'Ghastly at first, then improving.'

'Uninteresting.'

'Uninspiring.'

'All this is very, very good.'

'Penance for sins.'

'You will question your own sanity.'

Wainwright didn't really like the Pennine Way. And he prided himself on telling it like it is. But he was honest enough to give full measure, whether instructing a walker how to cross a misty fell, inveighing against the Ministry of Defence for uglifying the landscape, or allowing his inner romantic – a very sensitive and imperfectly hidden part of the Wainwright persona – free rein in contemplation of scenery that moved him. 'A charming riverside stroll,' he remarks of the Pennine Way as it heads northwest from Middleton-in-Teesdale, and that does it justice this morning. The Tees, swollen with yesterday's rain, comes charging over a shallow, rocky bed, its conversational rumble dulling most sounds in the narrowing valley. I follow it up past the cascades of Low Force to where the steady growl of the river changes to a pulsing roar. A flicker of white on a level with the path, and a fine mist over where the river should be. But the Tees has dropped away into space, a truly dramatic plunge of nearly a hundred feet, not a sheer fall but a majestic pouter-pigeon curve of solid brown water laced with white. It thunders into the rock basin below with a batter of sound that thuds like a waterwheel on the eardrums.

High Force is the biggest waterfall in England, says Wainwright. He means in terms of sheer volume, noise and spectacle, and with the Tees in spate like this he's absolutely spot on.

It is a dark volcanic rock called dolerite that forms this giant step in the bed of the Tees, as well as the columnar steps and cliffs beside the river. They are outcrops of a great subterranean tongue of dolerite known as the Whin Sill that licked its way in molten form across the north of England 300 million years ago. Where it squeezed between the layers of limestone here in Upper Teesdale, it baked the surrounding rock into sugar limestone, coarse stuff packed with minerals that makes ideal bedding for the most delicate of plants. And Upper Teesdale has those in spades – a relict flora of arctic-alpine plants that has clung on along the slopes and ledges of the valley since the last Ice Age. I'm so looking forward to seeing these tiny jewels of Teesdale today that I abandon the Pennine Way under Bracken Rigg, a couple of miles upriver of High Force, and follow a grassy old drover's road called the Green Trod up over Cronkley Fell. This bald green brow of hill rises abruptly from the west bank of the Tees, and I've had success before in finding arctic-alpine rarities along its broad nape.

A redshank flies across my path with a white flash of wings, sharply calling *tieu, tieu, tieu, tieu*. A lapwing flaps past from the direction of the river – *peew, peew, peeeiw!* – hoarser and more expressive than the redshank, a cry very much like a cat in distress. Both birds are scolding me away from their nests. A curlew goes staggering across the upper sky, shaken and shoved by a gusty wind that is rising in the north-west. It's a bit of a slog walking into the strong wind, but Cronkley Fell soon begins to yield its treasures. On the way up, in a crunchy swathe of sugar limestone next to an old lead-mine spoil heap, I spot tiny white stars of spring sandwort, the purple dots of the anthers showing brilliantly against the pure white of the plant's five blunt-tipped petals. Spring sandwort can tolerate what most plants can't, the toxic traces of lead in the soil. At the top of the fell is a line of large exclosures, areas from which sheep and rabbits and their efficiently snipping teeth are shut out by strong guard wire. Outside the exclosures the ground has been grazed bare; inside there is a low grassy sward, dotted with the flowers I've climbed here to see. Miniature Teesdale violets of deep pink-purple; the splendidly, royally blue trumpets of spring gentians, their lips turned outwards in a five-spoked wheel of petals; and my personal favourites, tiny bird's-eye primroses, each 'eye' egg-yolk yellow at the hub of petals that are as pink as an electric guitar or a starlet's fingernails.

These low-growing, exquisite little flowers caught John Hillaby's attention when he came this way in 1965. 'In this valley,' he wrote, 'a tundra has been marvellously preserved; the glint of colour, the reds, deep purples, and blues have the quality of Chartres glass.' And they do. The plants hug the ground modestly, tucking themselves down out of harm's way as their ancestors did to survive the freezing tundra winds after the last glaciation. It's only the peculiar make-up of the rocks and soil here, the fact that winter and rough weather tend to hang on in Upper Teesdale long after milder spring weather has penetrated the surrounding valleys, that inhibits competitor plants from migrating to the ledges and fell tops of Upper Teesdale, and crowding out these most senior and vulnerable of the dale's flowering inhabitants.

At seven in the evening, yawning like a horse, I lean out of the doorway of the Langdon Beck Hotel and question the north-western sky. Mackerels and mare's tails: it doesn't look good for tomorrow. And it isn't. Come the morning, the weather has reverted to winter. It's cold and windy, with the thermometer at 7°C and whitecaps on Cow Green reservoir. Long white banks of mist roll along Meldon Hill and Murton Fell, under which the Pennine Way makes a great westward swerve out of its generally northward course. That's where I'm headed, crossing ten miles of bleak and exposed moorland in order to drop down to Dufton village in the Eden Valley where Jane has offered to meet me. I think of the dismal day of wind and rain, the boggy slog that Dad and I endured on these moors when we made the crossing all those years ago, and I sigh with self-pity as I pull on thermals, fleece and thick winter coat, gloves, scarf and hat. Good Lord above, I thought I'd seen the last of all this stuff till November.

What's good for spring gentians, climate-wise, isn't necessarily so for walkers on the Pennine Way. But things turn out just fine. The first person I meet is the young farmer from lonely Birkdale, standing tall in the saddle of his quad like a circus rider as he guns the machine up the track. He is the picture of ruddy-faced health and wellbeing. He's in shirtsleeves, his only concession to the weather a beanie hat pulled low over the ears. He waves, his dog barks and they roar by, trailing a pong of engine smoke that soon dissipates in the wind. I stride on with proper vigour now, past the farm, over Grain Beck and on up a brand-new stone-surfaced track with the creamy falls of Maize Beck rustling in the valley below.

Dad and I had struggled across the misted moors of Rasp Hill at Alfred Wainwright's command, up to the shins in peat slutch, heads down into the wind and rain, querying the compass and quizzing the map. But today's crossing is – comparatively – a doddle. The murk is good enough to keep its distance. The stone farm track makes a clean and unmistakable path across the peat, and the turn-off for the crossing of Maize Beck, a spot where Dad had had to admit that we were lost – and yes, Alfred, where he did question his own sanity, and mine – is actually waymarked these days.

The weather must have cleared by the time father and son got to High Cup, the most famous viewpoint on the Pennine Way, because my notebook of the walk has a rough pencil sketch of a figure with legs that could easily be Dad's, lying back against a rucksack and savouring the prospect. A mile-high glacier gouged out this valley twenty thousand years ago. 'A fault valley with winding stream several hundred feet below,' say my notes from 1975. 'After the barren slopes, a sight of the lush Eden Valley with neat fields, trees. The valley sides just sweep in an unbroken arc down to the river. Grass and screes, then up symmetrically the other side. The Lake District as a blue backdrop over the far side of the Eden Valley.'

Nothing seems to have changed in any way. Those forty years have not even been an eye-blink to rock, grass and scree. I stand at the brink of High Cup and stare my fill. In a few minutes I'll jog on along the rim of the great cleft, then down into Dufton. I grin as I remember what befell Dad and me on the next day of our walk, a pea-soup mist that blanketed Cross Fell, birthplace of the River Tees, and beguiled us like a mischievous hob.

I become aware of a lithe young couple in Lycra running gear, bending and stretching just along the rocks. They are casting funny looks my way. Old fool, how did he get up here? Standing there, grinning to himself. I resist the temptation to make them a facetious bow. Have a nice day, y'all. I take a last look out over High Cup, and set off for Eden.

June

In June the man inside the man is
young and wants to lend a hand,
And grins at each newcomer.

Tɪɴɢᴡᴀʟʟ ᴀɪʀꜰɪᴇʟᴅ ᴡɪᴛʜ Jᴀɴᴇ on a cold, blowy Friday in June. Midsummer is two days away. Nothing much seems to have changed since 1989, when I last flew out of this Shetland Islands airstrip – a tiny hut, a DIY approach to baggage handling, and an informal attitude among the staff who stand watching a game show on the TV in the 'departure lounge' half of the hut as they wait for something to happen. As a portal to the otherworld of the island of Foula, Tingwall is perfect. They do check your ticket, just about, and they weigh you so that they can balance the tiny aeroplane that's about to carry you away. A mordant wisecrack, and that's your lot.

The Foula plane comes buzzing in. It is an Islander, a tiny pack-'em-in twin-prop plane with fixed undercarriage, giving it the rather charming aspect of permanently being about to sit down. The plane to Fair Isle that I'd taken last time out from Tingwall had been an Islander. This one seems exactly the same size and shape. The same can't be said of today's slim and trim pilot. The 1989 model had been a giant from the Outer Hebridean isle of Lewis, a humorous multi-stone maverick, absolutely determined to squeeze every last drop of fun out of his job. 'Come and sit up front with your camera,' he'd instructed me. 'We'll take a turn round Fair Isle, get you some good shots.' So he did, by banking like a Second World War fighter around the Pictish broch on Mousa. He followed that with a low-level swoop over Sumburgh air-port's runway that elicited gasps of consternation from the passengers behind, and a laconic comment in the headphones from the control tower: 'I hairrd that, Bob, but I saw nothing, ye'll be pleased tae know!'

Things are different today, but informality still seems the watchword at Tingwall. The airfield staff shove us and our bags into the little plane, the pilot gives a cheerful thumbs-up and pulls the Islander into the air after the briefest of run-ups. We wheel and bank. Foula lies ahead on the western horizon in evening sunshine, its profile of three giant steps facing north, the bulging hill of Da Noup making a hump to the south. Cumulonimbus clouds behind the island rise twisting to the north. We roar over geos, peat stacks and lumps of dour moorland a thousand feet below, then head out across a royal blue sea. Fifteen minutes of flight, and we are banking in front of Da Noup and sliding down to rumble to a stop along Foula's grav-elly landing strip.

The Shetland archipelago is as far north in Britain as you can go, and on summer nights the sky is always light there – the Shetlanders call it 'simmer dim', the dim light of summer. The island of Foula, twenty miles out in the Atlantic west of the archipelago, with its gigantic cliffs and twin mountain peaks, hundreds of thousands of sea-birds, marsh marigolds and powder-blue squill and floods of orchids, seems exactly the place to spend the longest day and shortest night of the year. Ever since hearing the Boys of the Lough play the leaping tune 'Da Shaalds o'

Foula' back in the 1970s, to me this remote Shetland out-post has represented a never-yet-visited Tír na nÓg. Foula has a human population of fewer than forty, I've heard, and these hardy crofters live and work to their own calendar, the ancient Julian one which celebrates Yule on 6 January and New Year's Day a week later. Over and above that amiable aberration, the Foula folk are legendarily hospitable and 'young inside'. So my Shetland spies report.

Fran Dyson Sutton, diminutive, endlessly positive and cheerful, is there to meet the plane along with partner Magnus and tiny son Alfie. Fran came to Foula as a fifteen-year-old on a working visit with the Brathay Exploration Trust, and fell in love with the island. She wrote two university dissertations on Foula, and kept returning until she saw the light and moved there permanently. Like all the islanders she multitasks, clean-ing and doing IT at the school, helping run the Foula Ranger service, and managing Ristie, the self-catering cottage up at the north end of Foula where we are staying. Half the islanders have turned out for the plane, to meet guests or friends, to collect cargo or to send things and people to the mainland. Kenny Gear, owner of Ristie and our next-door neighbour at Freyers, is there with his two tinies, Alma and wee Davie, who reaches out a hand to touch my beard with an incredulous expression.

The island has one strip of tarmac road which runs from the main settlement, Da Hametoon, north for three miles to Da Nort Toons where Ristie lies. Foula cars are mostly superannuated jalopies with residual number plates and 'relaxed' licences, one or more bald tyres, and nothing as technological as a handbrake. Fran's is a better class of wreck, and she drives us in it past the old school and the new one nearby, a fine Scandinavian-looking modern design with solar panels ranged along its fence, then the turn to the post office (no shop or pub on Foula). Another turn leads to the little harbour where the boats stand winched up on massive cradles out of the often turbulent water. Foula lacks a good sheltered harbour. 'We've had horrendous weather, rain and wind, until now,' says Fran, 'so you must have packed the good weather to bring over with you.'

Sunshine slides across the tall flank of Da Sneug, Foula's highest peak at 1,371 feet. The island's spine climbs from south-east to north-west, a sinuous backbone rising from the airstrip through Hamnafield and Tounafield to Da Sneug, then dipping to a saddle before rising again to the peak of Da Kame. Of all the cliffs in the British Isles, only Conachair on the island of St Kilda is higher than Da Kame, which falls twelve hundred feet sheer into the sea from the north-west shoulder of Foula. South of the island's ridge and parallel with it runs Da Daal, the Dale, a broad glacial valley; north of the ridge the landscape declines through a peaty and loch-spattered upland to a rocky north coast, beaten into coves by the sea.

We top a rise and look down on the northern extremities of Foula. Ristie and its neighbour Freyers are the two most northerly houses, and the remotest, on the island. Both are L-shaped, with a door in the southern angle of the walls to offer some shelter from the wind. Just offshore stands a great rock, Da Gaada Stack, its 130-foot summit slab supported by three separate columns hollowed out by the sea. The waves break over the top of Da Gaada Stack in the worst of the winter weather, says Fran, and seeing the wind-whipped agitation of the sea in midsummer we can well believe her.

As we look out west from Ristie's living-room window at 8 p.m. that evening, the sun is still high in the sky. There is a complete absence of trees – the constant blowing of salt-laden wind forbids them. Ridges and skylines run unbroken. We watch a snipe jumping vigorously up and down on the wet margins of a burn, pouncing as a cat would pounce on a mouse, driving its long beak into the wet earth to suck up morsels. A great skua ('bonxie' to Shetlanders) gets to its feet, stretches its white-tipped wings up and back as far as they will go, and opens its beak wide with neck craned up and forward, apparently yawning and stretching simultaneously – an odd, dog-like thing for a bird to do.

At nine o'clock we set out to walk in the simmer dim. The sun declines, framed in the grey stone arches of Da Gaada Stack. The soffit of the stack looks fractured and ready to fall, perhaps next winter, perhaps a century from now. The sea surges, thrashing at the base. Other stacks nearby, and the low cliffs to the east, are composed of black rock canted at 20 degrees and shot through with dull red volcanic dykes.

We follow the fields rising south-west to Da Est Hoevdi, the 'eastern headland', where the ridgeback of Soberlie forms a green wall overhead. The ridge ends abruptly in a vertical cliff that falls six hundred feet straight into the sea. As the sun sinks into a bank of cloud advancing from the west we start off up the flank of Soberlie. Three quarters of the way to the ridge, the slope has steepened to more than 45 degrees. The grass is skiddy with the evening's accumulation of dew. We won't make the top before dusk, and a slip on this slope would pitch us head-over-heels into the valley below. Reluctantly we zigzag down again.

On the way home, looking out west, we see the giant structure of an oil rig floating under the clouds, twinkling orange and yellow, as bright and unlikely against the darkening sea as a Hollywood spaceship. I recall the North Sea oil workers we saw at Aberdeen airport, heading for one of the big oil-industry helicopters and the flight out to a platform. They had barrelled past us at a fast stride, rolling along with the swagger of men in a closed, elite group – the army, perhaps. Solemn-faced, preoccupied, some shaven-headed, others with stubbly Jason Statham beards. You could smell the testosterone.

It's nearly midnight when we get back to Ristie. Tremendous snoring erupts on the far side of the flimsy door that separates our bedroom from the other half of the house. It doesn't keep Jane awake, but I decamp and make a bed on the broad windowsill in the living room. Looking out of the uncurtained picture window I see the black back of Soberlie cut high against a sky that is all wet cloud, but still light enough to delineate the hill – the strange effect of the simmer dim.

Next morning, lying in my windowsill throne, I watch a couple of rabbits at their grooming on the turf outside.

One is quickly finished and ready to play, but the other is far more scrupulous. It sits up and shakes the dew fastidiously from its paws, then licks each paw before passing them over its ears and face. Then it turns its attention to its flanks, chest, back and hind legs, stretching them out, licking and nibbling, giving its excretory organs a good cleaning. It is exactly the grooming a cat gives itself, with the same stops and looks around.

The playful rabbit stalks the ablutioner, using a big stone as cover. Suddenly it comes leaping over the stone and pounces. The grooming rabbit has appeared completely absorbed in its toilet, but it has been keeping one eye out, and dashes away two or three paces in the nick of time. Hunter and hunted remain frozen for five seconds or so; then the clean rabbit resumes its toilet, and the other takes up station behind the stone once more, its hiding place betrayed by the twitching black tips of its ears. Eventually something too high in the sky for me to see sends them scuttling off – probably a bonxie passing overhead. A bonxie can easily take on, and take out, a rabbit.

After breakfast I stand outside Freyers in the wind and chat with Davie Henry, cousin of Kenny and Bobby Gear. Davie is no stick-at-home. He has travelled all over the world. But he hasn't lost the islander's quiet, polite diffidence. Is there a chance of some music? 'Well, yes, I think the tune will be with you at Ristie tonight,' nods Davie.

The black and brown rabbits run across the peaty turf, and sheep with blunt primitive dog-like faces wander, trailing long skirts and scarves of wool. The fleeces of Shetland sheep are so loose that many sheep moult before they can be gathered and clipped. It's scarcely worth a crofter's while to shear the sheep; Shetland wool in its natural shades of dark brown, black and ginger can't easily be dyed, so is no use to the commercial garment industry. It is fabulously warm, though, and traditionally the fishermen wore it to keep out cold and wet. 'Heavy sweaters,' Davie says, 'with the motif of the island, and one for the family, too, worked into the pattern. That way, if a man was lost at sea, when he washed up he could be identified as to his place and his folk.'

Fran picks us up from Ristie in her rattly motor and drives us down to meet Sheila Gear at the post office. Sheila is the postmistress of Foula – also its chief historian, naturalist and folklorist. Her grandfather owned Foula. She spent much of her childhood out here, and loves every blade and stone. She married islandman Jim Gear in the 1960s, and brought up her children on Foula. Tough, resourceful, full of strength and energy, she has a restlessly enquiring mind and a deep and sad scepticism about the future of the planet, and of Foula's wildlife in particular.

The temperature is 7°C. Sheila walks in jeans and gumboots. I walk in thermal long johns and vest, fleece and thick winter coat, walking trousers and waterproof overtrousers, walking socks and boots, fingerless gloves, outer gloves, scarf and woolly hat. And I'm none too warm, at that. We set off up the broad south flank of Soberlie, with Blobers Burn gurgling in its peaty cleft to our left. 'The rocks in the burn look kind of blue,' Sheila says by way of explanation for the name. She bends and plucks a stem of soft rush, splitting the shaft with her thumbnail and teasing out a length of white pith. 'When it was dry, this was the wick for the rush light they used to use, floating it in a dish of fish oil or seal blubber. At first rush lights were in the house, and then as better lamps came in they would have been relegated to the byre.

'Do you know how to tell your rushes from your sedges? No? Roll the rush stem between your fingers. It's smooth and round, see? But here,' she stoops and pulls a dark green quill of sedge, 'feel the angle? There's a rhyme:

'*Sedges have edges,*
Rushes are round,
Grasses have joints
Right down to the ground.'

Climbing the upland slope, we learn the Foula names for wild flowers. 'Tormentil? We call that bark.' The grassy banks are layered with heath spotted-orchids and spattered with sky-blue dots of vernal squill. 'Grice's onions,' says Sheila. 'A grice is a pig, and they loved to root up the squill.' Sweet vernal grass whispers under our boots. 'See the feathery edges? That holds the scent that makes the hay palatable to sheep. Otherwise they'd be much less keen to try it. They have their own minds, sheep.'

Drops of Foula history trickle through our talk. The early nineteenth century, when the Scott family owned Foula, was a bad time for the islanders. Under the system known as debt-truck, all money had to be spent, all goods bartered at the island shop, a monopoly owned by the landlord. The men had to catch fish which the landlord would sell to set against what they owed him. They got into enormous piles of debt that they couldn't get out of. Arable land was divided up into equal plots and every-body worked their own; but every few years the ownership of the plots was rotated, so there was little incentive to make proper, permanent improvements to the land. The same went for the houses – families were expected to ren-ovate their dwellings, and then move on to a derelict one and do the same for that. So no one wanted to try too hard or invest too much handiwork in a house they wouldn't occupy long.

We pass a circular construction of stone, with stones piled on top to form a shallow conical roof. Below the roof stones, the interior is filled with a pale, dried substance. 'This is a mooldie koos,' Sheila tells us. 'People would pile the peat dust and rakings in here, and it would gradually settle to form a soft bed of peat mould. Then in the winter that'd be taken down to the byre and spread for animal bedding. It absorbed all the dung and urine, and the cattle wouldn't eat it, as they would any bedding of hay or straw. Then in the spring it would be spread on the fields for fertilizer.'

Overhead, skylarks scratch out their songs. Drifting veils of mist come and go on the back of the ridge con-necting Da Sneug with Da Kame, and the bonxies circle like hungry vultures. All the birds of hillside and coast have their Shetland names. 'Look, a wheatear,' says Sheila. 'We call that a stinkle. What else? Well—' she points to a fulmar planing past on stiff wings, 'there's a maali. The

Arctic skua, the one with the spiky tail and the white wing flashes – we call them skootie alan. Tammie noorie, the puffin, you probably know that one. A shag is a scarf. And, well, everyone who comes to Shetland in the summer gets to know the bonxie!'

Bonxies are big and burly, built and shaped with a dash of vulture. In the breeding season, with chicks to protect and territory to defend, these great skuas become nasty bastards – from a human perspective, anyway. Today they are trying to knock our blocks off. They bank and wheel, flying energetically away to gather momentum, then turning to launch low-level strikes on us, pulling up and away when only a foot or so from our faces. They honk and caw; clumsy, charmless birds, bandits, bullies of the air who chase and bump smaller birds till they disgorge their dinners. Usurpers of the rightful territories of Arctic skuas and gulls, murderers of red-throated diver chicks and puffins, we all agree. Then I almost tread on a pair of bonxie chicks, tiny balls of fluff the colour of milky coffee, who have ventured from the bare hollow of a nest their mother has trampled in the grass. They crawl away, then give up the struggle and lie blinking and wheezing up at us among the grasses. Aaaaah! Adorable! Who could hurt such a dear little thing?

Not so long ago, Foula's bonxies kept to the moorlands and hilltops. Few would have ventured far enough down the mountainsides to have bothered the red-throated divers on their lochs. Now the aggressive predators have waxed fat and successful. We hear of one resident who has made it their concern each spring to roam the island stamping on bonxie eggs. Thinking about the plight of the Arctic skuas that have been driven off their customary breeding territories on Foula by these muscular pirates, it's hard to know how to feel about the bonxies. Fish-breathed and bulky, unglamorously patchy of colour, dark and menacing on the wing, croaky of voice and murderous of habits, they are utterly unlovable in an anthropomorphic sense. But they are not supposed to be our cuddly friends or love-objects. They are wild birds, and – like the Arctic skua – in severe decline.

In spring and summer the UK hosts 60 per cent of the world's population of great skuas. They fly north from their African wintering grounds, and all but a handful settle to breed in the Northern Isles – Shetland and Orkney. When they are not robbing other birds, they tend to eat sand eels, and whatever trawlers throw back into the sea. In times of food shortage, they are not above snacking on bonxie chicks from neighbouring colonies. Bonxies suffer heat stress in warm weather, so climate change threatens them too. It might seem as though they have taken over on Foula, but appearances can be deceptive. In fact, their numbers have been falling over the past few decades. In 1980 there were about 2,800 pairs nesting on Foula, each pair successfully rearing, on average, one chick between them. By 2003 pair numbers had declined to 2,200, and successful chick production was down to one between two pairs. By 2010 a further fall saw 1,600 pairs producing fewer than one chick between two pairs. In 2012, three years before today's walk along the banks of Soberlie, only one chick fledged for every four adult pairs of great skuas. And that trend of lower breeding success has been echoed across Shetland, even though the bonxie colonies of Mousa, Noss, Fair Isle and Hermaness have been growing in number. The decline has been strongly linked to overfishing of sand eels, tiny fish that are scooped up in their millions to be turned into agricultural fertilizer and animal feed. When sand-eel fishing off eastern Scotland was banned in 2000, some seabird numbers rose. But fewer fish being caught means fewer being tossed back into the sea by the trawlermen. Deprived of this easy source of food to which they had become accustomed, the bonxies have turned increasingly to predating on other species, notably the Arctic skua on Foula, snatching their food, taking their chicks, pushing them out of the lower land on the island and driving down their numbers and breeding rates.

The Scottish Fishermen's Federation blames the decline of sand eels on the recovery in the number of predatory whitefish in the North Sea – cod, haddock, whiting – following temporary restrictions on fishing those species. And warmer winter temperatures may be having an effect, too, reducing the length of time that the sand eels spend hibernating in the sand without feeding, forcing them to 'wake up' and look for their zooplankton food at the time of year when it's least abundant. But sand-eel larvae, which hatch in March, are actually increasing in number. So nothing is easy to pin down as the root cause of the North Sea seabirds' decline. 'Wait and see' just about sums it up. So, more ominously, does 'Watch this space'.

*

Up at Summons Head we sit with our feet in a drain, a shallow trench dug right at the cliff edge to deter sheep from the abyss. From here you get a far better view of Da Kame than when you are standing at the summit and all the drama is out of sight below your boots. The almost sheer cliffs descend twelve hundred feet at an angle of 80 degrees or steeper, the apex in swirls of mist, the foot in a white collar of sea foam. Fulmars pack the ledges, while others spin slowly in the air currents far below like the first flakes of a snowfall.

A herd of tiny Foula ponies grazes nonchalantly at the cliff edge. Sheila approaches a couple – 'Horsie, horsie, horsieee! These are my daughter's, so they know me.' But the ponies turn tail and trot off. We climb in zigzags up the five-hundred-foot flank of Da Kame. Fulmars nesting in the rocks and bonxies on the steep grass slope are equally indignant at our approach, though none of the fulmars we disturb take the opportunity to spit bright orange fish soup at us, a disconcerting defensive reflex of theirs. Beside one bonxie nest we find the two splintered halves of an egg thieved from a guillemot nest, blue with black scribbles, smashed and sucked.

Up at the peak of Da Kame there is no indication, except for a slight upward incline towards the brink, that you are on the edge of such a mighty drop. Sheila leads us on down the far side to the headland of Nebbifield two hundred feet lower, from where we can see the cliff knifing down into the sea, fulmars in every crack and crevice, puffins nervously shifting back and forth at their burrow entrances on the grassy brink. To the south across a bay rise the cliffs of Waster Hoevdi, the 'west headland', a green tongue of land licking out into space. This was where the cattle were driven from the east side of Foula every spring to be pastured for the summer – a high grassy tableland falling sheer into the sea on all sides. The cattle were left unguarded, with occasional visits from their owners to make sure all was well. The cliffs of the headland are seamed with cracks where guillemots stand lined up shoulder to shoulder like commuters on a platform.

Down on Waster Hoevdi we look back at the cliffs of Nebbifield. Gannets with pointed black-tipped wings wheel far beneath us, yet far above the Scrodhurdins, the scree boulders that step seaward from the foot of the cliff. Sheila points to a steep rib of rock running down at 45 degrees to a tiny knob of grass fifty feet below us, a seat in the void. 'That's where I used to sit with my dog when I was a teenager. My parents never knew.' The nape of my neck crawls at the thought of it. We stare our fill, then climb back up beside Da Burn o Waster Hoevdi to the ridge between Da Kame and Da Sneug. Far below lie Da Lochs o da Fleck, a string of lochs, one shaped roughly in profile like Australia. Red-throated divers nest there, says Sheila, though she isn't sure if there have been any chicks this year.

Our homeward path lies along an invisible track, the Oxna Gaets, looking down on Da Lochs o da Fleck in their upland valley of blanket bog, set among brilliant green patches of sphagnum and tangled threads of sheep paths.

A snipe's nest lies in the grass on the Oxna Gaets, a single khaki-coloured egg in it. Three greylag geese pass three hundred feet below, their tails flashing white in the sun. Clouds move across Foula, trailing their lower skirts across the ridge on a level with us as we begin our descent.

It's a hundred years since the Oxna Gaets has been used for the purpose its name suggests – to bring the cattle up the 'gaet' or track and over the ridge to the summer pasture on Waster Hoevdi. We traverse the long northern slope of Da Sneug diagonally, trying to determine the path by unfocusing our gaze and not straining too hard to make it out. One way or another we make it down to the Burn o da Craig in the valley of Nedderafandal where red-stemmed marsh dandelions are growing. 'Rarer than hen's teeth,' Sheila tells us as we bid her goodbye on the road to Ristie.

Back at the house, there's work to be done. How many folk will turn up with their instruments and voices tonight, and when? We don't know, but we need to be good hosts, or hosts of some sort anyway. We inspect the box of groceries from Robinson & Morrison, Grocers, that the Foula boat deposited for us three days ago. Some tea, some coffee, a few biscuits, a bottle of red wine, a bottle of Jura whisky. It doesn't look much, even when spread out among cups, plates and glasses on the living-room table.

There's no need to fret, though. People don't start arriving until gone ten o'clock, and when they do they bring cake and biscuits and plastic bags of drink with them. Half the population of Foula crams into Ristie's living room. They fill the sofa, the chairs, the benches and the window seat. From under the table a selection of sawn-off log segments is rolled out to make more seats. Davie Henry perches near the big window with his mandolin. Sheila Gear's husband Jim carries in a mandolin and a guitar. Dave and Donna bring guitar and whistle. Sheila's daughter Penny brings three of her children. Ten-year-old Jack, Foula's only schoolchild, sings a funny song like a bird. His older brother Robert sits silent and shy until a guitar is lowered into his hands; then he picks out the chords to 'House Of The Rising Sun' and everyone does their best Eric Burdon impression. Johnny Cash and The Beatles pay us a visit, Ralph McTell and Bert Jansch are pulled forward and punished. But most of all there are tunes, starting with 'Da Tushkar', led with tremendous snap and life by Davie Henry. I find out, not wholly to my surprise, that he's been a touring musician for decades, a member of Shetland's home-grown band Hom Bru.

'Da Fields o' Foula,' says Davie, and up springs the tune I've known for ever as 'Da Shaalds o' Foula'. Fields o' Foula? So what are 'shaalds', then? It turns out that 'shaalds' are shoals or reefs, and Da Shaalds o' Foula is a particularly nasty specimen a couple of miles east of Foula. On 8 September 1914 the White Star liner *Oceanic* was wrecked there in calm weather and good visibility as the result of a navigational error, a huge embarrassment. With the First World War scarcely a month old, the cause of the disaster was hushed up in the interests of national prestige. Foula's shaalds have sharp teeth, and Foula wit is pretty sharp, too. 'Now,' says Davie, 'one on top of Da Shaalds o' Foula,' and he rattles out a jolly hornpipe called 'The Steamboat'.

Towards midnight I look beyond Kenny Gear's silhouette in the window, out to where Soberlie rises massive and black against the grey midnight sky of simmer dim. It's the moment and the setting for Foula's anthem. 'Da Song o da Papa Men' tells of the remarkable fishing expeditions the islandmen of Papa Stour would make in their six-oared sixareen fishing boats, rowing out into the Atlantic until Foula had slipped below the eastern horizon – 'rowing Foula down'. They would be guided home by 'da moder-dye', the mother wave, an east-flowing current discernible to practised eyes amid the tangled roads of the sea, or failing that, by the scent of flowers carried from their home island on the wind.

> Oot bewast da Horn o' Papa, rowin Foula doon.
> Ower a hidden piece o' waater, rowin Foula doon.
> Roond da boat da tide lumps are maakin,
> Sunlicht trow da clouds is braakin,
> We maun ging whaar fish is taakin –
> Rowin Foula doon.

We sing as we enter Midsummer Day. The deep chanted chorus has a melancholy cadence, like the haunting tune of 'The January Man'. I finger the guitar that has insinuated itself across my knee and run Dave Goulder's words for the month of June through my head, and I think of Kaveh Golestan and his 'man inside the man', the fearless war reporter I never knew, the teenage dreamer I loved who would have relished this moment to the hilt.

The guitar was the instrument that liberated my whole generation of would-be music-makers who never learned to decipher the dots. You could fail at playing the violin and the piano, you could fail at reading music, but all you required to play the guitar was the will to do it. You only needed to know three chords to be on your way. The guitar was versatile – you could hammer out a raucous Beatles or Stones tune at a party, or fingerpick something by Bob Dylan or the Incredible String Band to try and impress a girl. The guitar was portable, shaped for the knee, and tough enough to bash on a rock without dire harm. You could take a guitar down the road with you, as Kaveh and I did when we hitch-hiked to Istanbul in the summer of 1968.

If I had to choose one exemplar of the hippy dream in 1968, it would be Kaveh Golestan. We were attracted to one another like a pair of magnets when we met as teenagers in raffish, ramshackle Brighton. Kaveh was exotic, a boy from far-off Persia with the cheekiest of smiles, who wrote his own songs and was filled to the brim with peace and love. Long-haired, dreamy-eyed and fond of smoking red and black resins, he unquestioningly grinned at each newcomer. Kaveh was a year my junior, but several grades my superior at playing the guitar. I knew nothing about his background, nothing about his personal history. What did it matter? We were Donovan and Gypsy Dave, and once we'd read an article in a colour supplement about the overland route to Istanbul we couldn't wait to hit the hippy trail together.

In front of me now, very tattered and fragile, lies the map we used, with our route and overnight stops marked in biro. We set out from England on 14 July 1968, slept rough every night in deserted buildings or under trees and reached Istanbul on the 29th, so we made good time. We had £20, one loaf of bread, one pot of plum jam and one guitar between us. Romance and baseless optimism lifted us past hunger, boredom and some genuinely frightening moments. There was a lot of tension at the borders of Yugoslavia, Greece and Bulgaria, because the Prague Spring had only just happened. The authorities were nervous of long-haired youths like us, and their minions were correspondingly aggressive.

The Ostend police bawled us out when they caught us sleeping on a building site. The Frankfurt police bawled us out and relieved us of a few marks when they caught us hitching on the autobahn. We were dropped off by a tiny Fiat 500 in a midnight snowstorm at some high pass in Switzerland because we smelt too bad to be tolerated any longer at close quarters. Waking next morning under a cowshed warmed by the cows, and looking out on the snowy Alps, was the purest magic I'd yet experienced. We survived a tyre blowout and screaming skid in Italy, and an attempted kidnap and robbery by a lorry driver in Yugoslavia. The Skopje police bawled us out for looking unsavoury. We sang for our supper and rooftop lodging in northern Greece. The border police at Edirne gave us a slap for being cheeky. We reached Istanbul in fine style, in a gigantic American roadster driven at great speed by a drug dealer in a double-breasted blazer and white trousers with knife-edge creases.

We spent a fortnight high as kites in Istanbul, sleeping on the flat roof of the notorious Gulhane Hotel and eating cheap bowls of semolina and Nesquik at the Chocolate Pudding Shop, a famous hippy hangout. We talked nonsense by the square yard; we smoked and sang till our throats were raw, the guitar passing back and forth between us. We saw almost nothing of the historic glories of the city. I felt closer to Kaveh than a friend, closer than family. Then one morning my romantic brother of the road announced that he was leaving me and travelling onward to his family home in Tehran. That bolt came out of the blue, though it was perfectly in accord with the groovy philosophy of going where the wind should happen to blow you. I gasped, I regrouped, and I made my way back to England on my own, losing my passport and all my remaining money along the way. The great hippy adventure ended with a humiliating interview and the issue of temporary documents at the British Embassy in Brussels, a thorough and to my mind rather gloating strip search at Dover, and a limp homecoming, hungry and stinking and secretly glad to be back in one piece.

For nearly fifty years Kaveh remained at the back of the memory drawer. He would come to mind once in a while at the mention of Istanbul or a snatch of a Donovan song. The thought of him invariably brought a smile and a lift of the heart. Then, browsing the internet, his picture leaped out at me – a black-and-white photo-booth snap with his sister from the 1960s, Kaveh barely suppressing a cheeky grin. I saw his name alongside with a thrill of recognition, turned eagerly to the printed paragraph under it, and had the smile wiped from my face. Kaveh was dead. He had been dead for ten years, killed by a landmine in Iraq in 2003. Unknown to me, he had become a famous war photo-journalist across the Middle East, a chronicler of humble lives and of great changes in his native land. There had been outpourings of grief, expressions of loss from eminent names. The epitaph inscribed on his grave memorial in Afjeh near Tehran declared simply: 'He was killed while documenting the truth'. And the toll of recording that truth had been stamped on the face that looked out at me – not the grinning boy in the booth, but a solemn-looking man in middle age with a great snow-streaked peasant's moustache. Eyes that had seen too much, but were wide open for more, stared out from dark hollows under shaggy black brows. He looked far older than his fifty-two years; far older than I had looked at that age. Doing great things, he had become a great man: a humble and a gentle one, too. I felt unreasonably bereft at the death of this unknown person. It took me some time to understand that, quite as much as for Kaveh Golestan, I had fallen into mourning for my own young self.

All that I knew of Kaveh, I knew when he was scarcely formed, before the man inside the man took proper shape. Despite my current knowledge of his later life and death, he somehow still remains the same carefree lad inside my skull. But these images we carry are not always those that are fixed in golden youth. When I found a photograph in my father's dressing room not long after his death, it was a sad shock to discover that, in Dad's case, I had very little memory of the 'man inside the man', the man who could feel young and grin from ear to ear.

This dressing room of Dad's, smaller and pokier than the one that had doubled as my childhood bedroom, was in the house that my parents moved to from Hoefield House after Dad retired. It was a room I'd seldom been into before the last few months of Dad's life. He'd needed help at that stage to stand and wash at the little cold basin, all he'd wanted by way of 'en-suite facilities'. Next to the basin was a chest of drawers which he had packed with incomprehensible papers he'd thought would help his executors, meticulously stored in envelopes, each one labelled in his tiny slanted script. The short drawers were full of the chaos of his small belongings – empty hearing-aid boxes, wafer-thin lawn handkerchiefs, tie clips unused since the 1950s, a little heap of faded blue paper packets still holding the rust-eaten blades of King C. Gillette, T.M. REG. U.S. PAT. OFF. The top of the chest held a scatter of photographs, and I found the snapshot by chance among them. I took it to the better light at the window, and found that it was one I'd never seen before.

In the faded colour of the photo my Aunt Rachel, Dad's sister, sat upright on a sofa in my parents' house. Tall curtains had been drawn behind her, so it was night-time. She had a smart frock on. She looked as though she was on her third G&T. Something had just appealed to her wickedly salty sense of fun – that bold grin of hers was instantly recognizable to me. But who was this youthful, cheerful man sitting alongside, his arm around her, a huge uninhibited smile across his face?

I didn't recognize Dad at all, though the context of the photo said it must be him. I couldn't remember ever having seen him smile like that. All my lifetime's memories had been usurped by the man in old age whom I had only just relinquished to death. This in the photo was a different man entirely. He could have been in his very early sixties – not much older than I was today as I stood by the dressing-room window and studied the picture – but he looked ten years younger than that. If I'd got his age right, he must have just retired from GCHQ. That would have caused a few years to roll off him. He'd got a suit and tie on. He looked happy and relaxed, and he'd obviously had a glass or two – maybe they had just come in after a party or an evening at the theatre. And he was bubbling with suppressed glee, grinning with his sister as, Rach had often let slip, they used to grin and chortle over rude sayings when they were young.

> *Ma's out, Pa's out,*
> *Let's talk dirt –*
> *Pee, piddle, piss, po,*
> *Bum, fart, drawers.*

I was swept into an eddy of memories. Dad sending away to Ellisdons Joke Shop (Dept. 6) for rubber tramp's boots with bloody bare toes to wear to a fancy dress party, the joke redoubled when two left feet arrived in the post. Rolling back the worn green carpet in the dining room and dancing Sir Roger de Coverley with a bunch of friends. Laughing so hard he had to sit down when Mum crashed her toboggan on Bredon Hill and ended up in a snowdrift with her skirt over her head.

That was the man inside the man. And here he was again in my hand, a man I could have been friends with at fifty, a man feeling young and primed for fun. If I ever knew that man, I had forgotten him until this moment.

Around half past one in the morning the company in Ristie's living room make a move to go. Their handshakes and formal 'thanks for having us' outweigh our own stammered appreciation of the evening's fun.

On Midsummer Day we are up with the sluggard. The bottle of Jura, still half full, stands on the table among dreg-swilling teacups, biscuit crumbs and an unopened six-pack of lager. We need a good bracing breath of air, and Foula is exactly the place to provide it. This morning, the peak of summer, the thermometer still shows 7°C. The wind has veered round from west to east and is gusting fiercely, slicing a wind chill factor of several degrees off the temperature. We put on the same midwinter gear as for yesterday's expedition, and are not at all too warm.

We waddle out to walk the island road. Near the airstrip, Arctic skuas flap and screech. A snipe goes skidding across the sky in a frenzy of display, switching direction in a flash, diving with a sharp, prolonged hooting noise, the sound of the wind drumming in tail feathers held stiffly out sideways. Another swoops up and down in extravagant rollercoaster curves, flipping over on its back to show off a white belly. A territorial declaration, a piece of swagger for the females, or a symptom of distress at two swathed humans too close to its nest?

Several high, wailing calls, far louder and more penetrative than the gibbering of the snipe, echo off the slopes that rise to the upland plateau of Ouvrafandal. We keep our eyes peeled for whatever it is that's making these noises, which emanate from where the triangular Sandvatten or Sand Loch lies hidden beyond a fold of ground. As we near the turning for Ristie a sliver of the loch comes into view, and a smear of movement on the water catches my eye. We cross the fence and move cautiously towards Sand Loch, keeping our heads down and creeping as quietly as we can, until we have rounded a hummock of bogland and can get a proper sighting of the loch. A magnificent pair of red-throated divers is cruising there. Two dark fluffy chicks accompany them. Seen through binoculars at three hundred yards, the scarlet throats of the adults are just discernible in this dull northern light against their grey-white breasts. Their sleek heads with the long slim bills are turned to the sky, swinging ceaselessly to and fro in anxious regard of the bonxies that swoop and clatter over the loch. A bonxie can take out a diver chick just as easily as a young rabbit. What beautiful birds these adult divers are. They hold their heads proudly, dagger bills raised, the backs of their necks striped vertically in black and white as though neatly raked with a comb. Once again an anthropomorphic indignation takes hold of me. How dare the bonxies threaten this handsome pair and their dear little chicks!

With the snipe drumming and zigzagging overhead, we turn back to Ristie, where Kenny Gear has deposited a great shank of Foula lamb in the fridge. We cook it with potatoes, carrots and onions we've brought to the island with us, and spend the rest of the day licking grease off our fingers and burping contentedly.

Next morning we drag our suitcase and couple of backpacks to the door and load them into Fran's truck. The wind has swung round like a teetotum during our stay on Foula – westerly for the first two days, easterly on Sunday. Now it blows hard from the north, a grey gauzy wind laden with sea spray. Down at the airfield half a dozen islanders stand hunched over cigarettes in the lee of the shed. Young Jack wears a Fair Isle Bird Observatory hat ('Boo! Traitor!' someone jokes) and a huge grin. He's here to welcome a group of schoolchildren who are coming in for a week's visit from other tiny off-islands in the Shetland archipelago. It's an important moment for Jack and the other small-island children; none of them has as much contact as they or their parents would like with the wider world of Shetland youngsters. Jack will be leaving Foula to attend Anderson High School in Shetland's capital town of Lerwick next year, staying away from his home island for weeks at a time. Luckily he's a gregarious lad, but he still needs all the practice he can get in socializing with people of his own age. To put it mildly, the big school can be a tough place for children from the smaller islands.

The Islander is seen as a speck, low above the sea in the east. Jack and Fran unfurl a home-made welcome banner and brace it against the gale. The little plane roars in, shaken by gusts of wind, and lands in a scuff of gravel. Five or six children come bobbing out. They run towards Jack, hopping and squealing, 'We're on Foula! We're on Foula!' They are going to have the time of their lives, cold and wind notwithstanding.

We pack into the Islander, the pilot gives the thumbs-up and revs the engines. A last wave to Jack and Fran, and we are zooming up across the rocks and the angry sea into the grey buffeting sky, with Foula already dissolving behind us into haze and spray.

July

And in July the man in cotton shirt,
he sits and thinks on being idle . . .

I STARE UP THROUGH binoculars at the cowled face of the nun on the dusky red wall of Melrose Abbey, and she stares right back. Seven hundred years ago an anonymous mason gave the corners of her sandstone lips an upward curl, and a little hint of mischief still comes through to me despite the centuries of weathering that have evened out much of her expression. It's a pleasing image to carry away under the brisk skies of July as I head out of town along the well-trodden track of St Cuthbert's Way. This modern-day pilgrim path starts at Melrose in the Scottish Borders, where St Cuthbert entered the monastery in AD 651 as a shy shepherd lad, and runs sixty miles east to Lindisfarne or Holy Island off the Northumbrian coast where the saint continued his ministry before retreating to the Farne Islands, the bleakest shelves of rock he could find in the wild North Sea.

Cuthbert has been a favourite with me since my three-year stint at Durham University in the dog days of the 1960s. I know I was lucky to be taken on by the not-quite-college known as St Cuthbert's Society, almost entirely on the strength of a casual chat about science fiction at an interview with the Principal, Leslie Brooks. At Cuth's I found a collection of oddballs and queer fish, unclassifi-ables who couldn't be slotted into colleges with a more conventional ethos. To say the least of it, I wasn't a serious-minded youth. As a member of St Cuthbert's Society I did a great deal more laughing, quaffing and chaffing than sensible work. That's probably why I took away from Durham a very poor degree, along with the notion of Cuthbert as a rather more agreeable saint than most.

Actually, by any unbiased reckoning, the hermit who battled the demons of the Farne Islands was far from being a jolly soul. Modern psychiatry would probably label Cuthbert a sociopath suffering from paranoid delusions. But there was a time in his life when he must have been a lot more approachable. He gave up shepherding and joined Abbot Eata's religious community at Melrose at the age of seventeen, having had a vision of angels descending to earth on a beam of light to fetch St Aidan of Lindisfarne up to heaven. And Cuthbert soon became a restless traveller in the service of the Lord, earning the nickname 'The Fire in the North' as he criss-crossed the Cheviot Hills and their northern lowlands, burning with youthful zeal and baptizing heathen locals by the armful.

I head uphill out of Melrose on this July afternoon in a burst of energy, looking forward to following the path of the fiery young shepherd saint for a few days through the border country he galvanized so long ago. There's an added fillip, too, in being seen off on my journey by the sandstone nun on the abbey wall and her knowing little smile. And St Cuthbert's Way enjoys a satisfyingly direct start, cantering straight up the steep slopes of the Eildon Hills, a trio of small but shapely humps that crowd Melrose on its southern side. Runners from the town chase each other across the heathery flanks of the Eildons, families go picnicking among their summit rocks. The hills are Melrose's much-loved familiars, and the path that traverses them is beautifully waymarked and mown. It carries me up through gorse and ling to the saddle between Eildon Hill North and Eildon Mid Hill. A stony scramble, and I pop out at the top by the topograph on Mid Hill, a thousand feet above the town, looking down to the lacy stone arches of Melrose Abbey lit by the sun against low whaleback hills to the north. A cold wind blows boisterously from the north-west where the sharp little peaks of the Moorfoot Hills stand two-dimensional and blue. In the other direction grey shoals of cloud jostle like plump fish towards a line of porpoise-like hills rising pink and purple on the horizon – the Cheviots, waiting for my bootprints three days from now.

The three abrupt lumps of the Eildon Hills, upstanding in a low and gentle landscape, are the obvious place to build structures that need to be seen from afar. They have had their share over the years – an ancient burial cairn and a topograph on Mid Hill; a massive loop of ramparts, hundreds of Bronze Age hut platforms, and the foundations of a Roman signal tower on Eildon Hill North. Such hills attract stories, too, like iron filings to a magnet. King Arthur and his knights lie asleep under the hills, which are known to be hollow. There are legends of maidens gone astray, rumours of a great treasure buried deep. But the Eildon tale that outdoes all others is an early medieval ballad, the stranger than strange odyssey of Thomas the Rhymer.

Thomas, a poetical youth from Earlston up the Leader Water, is mooning in the Eildon Hills one day when he meets a fair lady on a milk-white mare. She wears a green silk tunic and a velvet mantle, and is seated on a saddle of ivory fretted with gold. Silver bells are plaited in her horse's mane. She has a bow in her hand, arrows in her belt and hounds on a leash. Thomas takes her for the Queen of Heaven – or at least, is polite enough to tell her that he does. She says no, she is but the Queen of fair Elfland.

She is a huntress, clearly, and Thomas is her prey today. She bids him kiss her, and when he does he passes into her power. The radiant queen becomes a toothless, bearded hag and gallops off with the poet to Elfland. It is a terrifying three-day journey in the dark through seas and rivers and torrents of blood, but at last they win through to a green garden. Here the seductress resumes her youth and beauty. She plucks Thomas an apple, saying that it will give him a tongue that cannot lie, and will also grant him the powers of prophecy. He is bidden not to breathe a word about the kisses they have shared – otherwise he will never look on the Eildon Hills again. Then they ride on to a castle where the lady's husband, the King of Elfland, holds court.

The King receives Thomas kindly. He is given a suit of elven clothes and a pair of green velvet shoes. Then the Rhymer is free to join the feasting and revelry in the castle. After what he takes to be an hour or so, the lady draws him aside and informs him that seven years have passed. 'Thomas, Thomas,' she says, 'the Devil is on his way to the castle to exact tribute, and he's sure to choose so pretty a youth as you. Quickly now, mount the mare, and I will take you back to the mortal world.' They gallop once more through the dark and the seas of blood, and Thomas finds himself back on the Eildon hillside. He has become older and wiser, and as the lady promised, he is possessed of the gift, or curse, of being able to tell nothing but the truth, and to speak of nothing but what is to come.

Thomas the Rhymer's brush with the Otherworld appears to be at an end. But fate holds a final twist for him. One day he is entertaining friends up in Earlston tower when a servant rushes in, panting and mazed. 'Sir, a hart and a hind have been seen outside, walking the streets of Earlston without fear.' Thomas is struck with beguilement. He runs down into the street, follows the enchanted creatures into the forest, and is seen no more by mortal eyes.

What a skein of tales and legends is tangled up in the tale of Thomas the Rhymer. Adam and Eve in the Garden of Eden. The Quest for the Holy Grail. Persephone abducted by Pluto, Ariadne abandoned by Theseus. The seducer Tam Lin transforming himself to test the constancy of lovely Janet. Fionn mac Cumhaill by the Lake of Sorrows on Slieve Gullion, watching in horror as the Cailleach Beara transmogrifies from a youthful beauty to an ancient hag. Young Rip van Winkle returning home an old man, to find time flown. Keats's knight-at-arms awakening on a cold hillside after kissing La Belle Dame sans Merci. The ballad is a rich allegorical stew of abduction myths, a compendium of wonder and paranoia filtered through guilt and shame, with a teasingly open ending. Is there even a moral, apart from 'Don't kiss and tell'?

*

Mid Hill's topograph is so eroded by decades of fingers rubbing its bronze face that most of the inscriptions are illegible. But I make out the name of Yeavering Bell, the Cheviot height I know best. I squint in the direction indicated and salute its tiny, far-off dome before skeltering back down to the saddle and on along St Cuthbert's Way. On a common bright with buttercups and spotted orchids I meet an elderly walker. 'This used to be all left uncut and neglected,' he ruminates, 'and you never saw an orchid here. But then it was left in a bequest to someone who cared about such things. Now the students from the agriculture college cut it every spring' – he throws his hands wide, indicating the wild-flower carpet – 'and look at what we get!'

Beyond Newtown St Boswells the Way cuts through a tanglewood to a sandstone cliff that overlooks the River Tweed. The river courses round a great bend, running quick and shallow over long grey ridges of shillets. As I walk the bank towards my night's lodgings I look out for Dryburgh Abbey across the river, but it is hidden away behind a screen of trees. 'A shame that you can't see it from here,' I'm told by a man with stick and dog. 'But you never could. I know this path pretty well; I've been up and down it for over fifty years, learning about the plants. That's my particular interest, yes. I used to cover this area of Roxburghshire . . . Don't you like the old names for the counties? We used to be Roxburghshire, bounded by Selkirkshire to the north and Berwickshire to the east, but it's all just Borders now. What was I saying? Yes, I used to cover this part of Roxburghshire for the Botanical Survey of Great Britain.' This is said with the old-school modesty of someone who would rather die than admit to any personal distinction. He has a quiet, definitive answer for any query. 'That yellow daisy-ish flower? Leopardsbane. Campanula is the one with the blue bell. And here – now here is a bush of philadelphus, mock orange, that I've watched growing all this time. This is the first time I've seen it for several years, so I'm glad to see it spreading.' He caresses the white flowers and spiky green sepals with wrinkled fingers, and smiles at the bush with the pleasure of meeting an old friend who's doing unexpectedly well.

Next morning, early sunlight on Dryburgh Abbey catches two memorials lying side by side – a simple monumental marble tomb under the arch of the north transept, and a plain soldier's grave in the aisle alongside. The marble slab conceals all that is left of Sir Walter Scott, the nineteenth century's most potent romanticizer of Scots history. Under the sandstone marker lies the Great War's chief architect of triumph and slaughter, Field Marshal Earl Haig. I could stand there and expend a lot of sententious thought on this curious pairing, but Dryburgh Abbey on a gorgeous July morning is stealing me away. I finger the dogtooth carving of a doorway and lay my cheek to the sunny warmth of cut sandstone. The perfection of the rose window above the cloisters, a stone flower outlined on the blue sky, catches the eye and holds it. Edward II looted and burned Dryburgh Abbey in 1322, Richard II in 1386. Fire destroyed the place in 1443. The

Earl of Hertford burned it twice, in 1544 and again the following year. How could such a fragile thing as a window of gossamer stonework have survived for so many centuries in this borderland where castles and churches, city strongholds and town walls were brought low in their hundreds by relentless fighting and unmitigated bigotry?

The night stair rises to the dormitory where the monks of Dryburgh shivered the long Scottish winters away in their sheetless beds. I duck into the vaulted chapter house. The faintest streaks of black and red paint still outline fresco traceries on a screed smoothed by a plasterer seven hundred years ago. Here is the hint of a ridge he left with the edge of his trowel; shoddy workmanship for a purist, if he was a purist and not a local jobbing jack of all trades hired cheap for the sake of saving a couple of pennies for the foreman's purse. As always when contemplating the scratches and furrows our ancestors left stamped on the work they made, I am brought up short and made to wonder about this unknown craftsman and his slightly unsteady hand.

Dere Street leaves St Boswells like an arrow, shooting south-east across the map as straight as only the Romans could cut a road. I get onto the old track by the tollhouse at Hiltonshill, and follow it for miles. Dere Street sets the pattern for St Cuthbert's Way today – a path running between hedges that form a screen through which the landscape shows in glimpses. The countryside declines south towards Teviotdale in a series of shallow valleys running east to west, and Dere Street hurdles them effortlessly in a succession of gentle rollercoasters. Roman roadmakers didn't worry about ups and downs. They just went where they were going, hard and straight.

A wren is singing its heart out in a stack of woodyard logs as I pass by. There are pale orchids among the trees and deep pink dog roses hung up in their branches. When I get to the hollow sandstone monument that marks Fair Maid Lilliard's grave, I find it full of blackberries, raspberries and tall Scotch thistles. The memorial slab fixed to the back wall of the tomb is blotched with lichen and badly eroded, but I can just about make out its rhyming lines:

Fair Maid Lilliard
Lies under this stane
Little was her stature
But muckle was her fame
Upon the English loons
She laid monie thumps
And when her legs were cuttit off
She fought upon her stumps

It's a great bash-the-English tale, that of Fair Maid Lilliard. And the Battle of Ancrum Moor, where the Scots under Red Douglas thrashed Henry VIII's forces on 27 February 1545, was real enough. Henry was looking for a suitable bride for his seven-year-old son and heir Edward, and had decided on the two-year-old Mary Stuart, infant Queen of Scots. But the Scots had other ideas. They rejected the whole notion, triggering a series of punitive English raids that became known as the 'Rough Wooing'. The English got Ancrum Moor badly wrong; out of an army of five thousand they lost one-third – one thousand captured, eight hundred left dead on the field. The slain included the two English leaders, Sir Ralph Evers and Sir Brian Latoun. Evers in particular had earned the name of a brutal man, and it only inflated Scots pride when the story got about that he had been bested by a young woman. Lilliard was from the nearby village of Maxton. Her family and her lover had been killed by the English, it was said, and she had rushed to the battle on the moor to exact personal revenge.

Does it matter that the tale was triggered by a romantic eighteenth-century clergyman, the Revd Milne of Melrose, later embellished by Sir Walter Scott, and is a mix of juicy details from the fourteenth-century Battle of Otterburn as well as Ancrum Moor? Or that the name of 'Lilliard's Grave' had its origin in the ruins of Lilliot Cross, a stone erected by monks from Melrose to guide travellers three hundred years before Sir Ralph Evers met his bloody end? Standing by the 'tomb' and looking back to the three humps of the Eildon Hills, I picture Lilliard fighting on her stumps. Another image comes to mind – the Black Knight in *Monty Python and the Holy Grail*, spitting out defiance as King Arthur lops off his arms and legs one by one. It's hard to take Fair Maid Lilliard seriously. But the eight hundred English soldiers and the several dozen Scots brained and crushed and choked in their own blood in this place on that February day – that is not so funny. For all the present peace and beauty of the Scottish borders, any walk through these bitterly contested lands is a wade through spilt blood.

The Cheviot Hills lie temptingly ahead, a lumpy line on the eastern horizon. St Cuthbert's Way runs through green and gold wheatfields squared off by hedges where yellowhammers sit wheezing for 'a-little-bit-of-bread-and-no-*cheeese*!' Dere Street is shaded by an avenue of silver birch and contorted old beech trees, footed in an ancient hedgebank that runs as straight as a die. The tiny animated blob of a walker appears at least a mile ahead, and grows gradually larger and better defined until we meet in a grassy hollow swirling with Scotch argus butterflies. 'Doing Land's End to John o'Groats,' says Penny from Ipswich, a sturdy figure under a towering pack. Seven weeks on the march, four or five to go. She's had just one blister, she proudly declares, 'second day out, and that was my fault. Sand in my sock, and I didn't stop to get it out.' The scariest bit? 'Oh,' she shudders, 'the Pennine Way over Cross Fell, without a doubt. The mist was so thick I couldn't see my own hand. A bit lonely up there!'

I smile at that, and Penny looks quizzically at me. 'Don't suppose you've been on Cross Fell in a mist?'

30 July 1975. 'Morning, lads,' said Mr Morris to Dad and me as he brought our bacon and eggs into the dining room at Sunnyside. He jerked his thumb in the direction of the window. 'Not looking too clever today.' It wasn't. All the fells behind Dufton had been cut off at the knee as though by a vindictive giant. Dad and I looked out on a

two-tone world: green down here, dense and dirty white up there. 'Ah, well,' said our host, 'at least we haven't the Helm today, eh?'

True enough, we hadn't the Helm, the vicious local wind that can rise out of nowhere to blast the top of Cross Fell and scatter the roof slates down in Dufton. But that was cold comfort. We'd limped into the village last night after one of those Pennine Way crossings you want to forget, a dozen rain-sodden, peat-splattered miles from Langdon Beck. A subtle bog above Maize Beck had lured us in. The gritty peat slutch with which it filled our boots had scoured our ankles into holes on the long descent from High Cup. We weren't in great shape for one of the Pennine Way's toughest days, the crossing of its highest point, Cross Fell, on a section supposedly sixteen miles long that would finish in the village of Garrigill. Hill mist of today's sort, so thick you could part it with your hand in front of your face, was just exactly what we didn't want. But we did have Wainwright's *Pennine Way Companion* for a comfort. The Master's obsessively accurate instructions and sardonic encouragement had seen us safely this far. Today would be 'a pleasant ramble' with 'fine views', apparently, followed by a 'rapid descent' on an 'excellent track'. That was all right, then. And at least we hadn't the Helm.

The Pennine Way adventure, the first long walk Dad and I had done together, was going well enough. We'd had a couple of pints and a chat about family things last night in the Stag Inn on Dufton village green. Dad wasn't used to pubs, and was wary of the beery good fellowship of taproom strangers. But the knowledgeable and purely local talk of a couple of shepherds and a lorry driver, into which we were soon invited ('Now then, lads, doing that Pennine Way, are you?'), had put him more or less at his ease. He still wasn't prepared to divulge a thing to any member of his family about GCHQ or his working life there, however.

Even from my very uninformed perspective at that time, 1975 felt like a dodgy year. America had just been driven out of Vietnam, tail between legs. Communism in extreme forms was rampant, with the Pathet Lao taking control in Laos and the Khmer Rouge in Cambodia, as well as Marxist governments establishing themselves in Angola and Mozambique. The good guys seemed to be on the back foot, though John le Carré in *Tinker Tailor Soldier Spy* was nudging us to question exactly who the good guys were. Dad loved John le Carré. But he didn't love the trade union officials – some of them Communist sympathizers, to say the least – with whom it was his duty to negotiate on GCHQ pay and conditions. I knew that much, from a couple of harrumphs he'd let fall. What I didn't know then was that he was desperately worried about the possibility of a strike that might cripple GCHQ's cryptanalytical computers, causing them to miss vital intelligence, and upsetting the delicate relationship of 'the office' with the Americans' National Security Agency.

In a little over two years, the guillotine would fall on Dad's working life, on all that incisiveness and decision-making. At exactly sixty years old, he'd be out to grass. On with the slippers, up with the feet? Not Dad. He'd be scrabbling round for something to do, some way to contribute. They would almost kill him with frustration, those sun-kissed uplands of retirement. He didn't want to sit and think on being idle – not at all. I don't think Dad ever read Jerome K. Jerome's *The Idle Thoughts of an Idle Fellow*, but how he would have agreed with the author: 'It is impossible to enjoy idling thoroughly unless one has plenty of work to do. There is no fun in doing nothing when you have nothing to do. Wasting time is merely an occupation then, and a most exhausting one. Idleness, like kisses, to be sweet must be stolen.'

Dad's working life was precious to him, a means of doing the right thing and justifying his existence on this earth. But at this late stage of his career it was draining him dry. He needed to steal a little idleness for himself, but he was morally and physically incapable of just putting his feet up or pottering about. If he were to allow himself to be idle, it would have to be productive idleness, idleness with an outcome. So he chose the punishing challenge of the Pennine Way, so tempting as an exercise in escape for a solitary walker with a variety of monkeys on his back. The fact that Dad then elected to do it in the company of a son in many ways unserious and immature, with a propensity for chatter and spinning of fantasies and drinking beer with strangers, says a lot about his undemonstrative love that I have only slowly come to understand since then.

Three Sunderland boys we'd christened the Greyhounds were staying at Sunnyside too. They were doing the Pennine Way in heroic stages, forty miles a day. We watched the Greyhounds sprinting off towards Cross Fell as we were lacing our boots on the village green. They were up in the mist and out of sight by the time we hit the walled lane to the hills. We climbed a steep and interminable roadway of potholes, blanketed in mist, that might well have reached halfway to the moon by the time it deposited us two thousand feet higher beside the radar station masts on Great Dun Fell. 'A monstrous miscellany of paraphernalia . . . grotesque contraptions,' spluttered Wainwright. 'Quite the ugliest of all summits.' But where were the contraptions, the lattice signal masts several hundred feet tall that ringed the fell top? The mist had drawn in so thickly that I almost walked into one of the metal legs before I spotted it. There was no seeing anything at all, certainly not the lie of the path, let alone the 'fine views'. It began to rain. What now? Turn back down out of the murk and make our way round to Garrigill by road? Impossible – it would mean a thirty-mile day. Carry on and trust to map, compass and the Master, Alfred Wainwright? 'Only thing to do,' said Dad, decisively. 'No good just standing here getting wet. Come on.'

He started off, Wainwright in hand, and was instantly swallowed in the murk. I knew he didn't have much of a clue where he was going. He just needed to go, somewhere, right this minute. I didn't want to follow him, but even less did I want to lose sight of him. So I plunged after him, following the orange blur of his backpack, as I'd plunged after his blue bathing trunks into the sea while learning to swim on holiday at Abersoch twenty years before. Back then I'd run out of the shallows in a panic after a moment's freezing immersion, but there was no running out of this. Ahead we could see perhaps five yards of a faint brown indentation in the grass that indicated the path. It dipped downward, met a squashy patch of bog and vanished. Nothing ahead, to left or right.

'Bugger it!' Dad fumbled out his reading specs. He brought the *Pennine Way Companion* close to his nose. '"The walking is very good, on dry firm turf." What balls!' We took a few steps forward into the mist. Against the laws of probability, it seemed to be growing even thicker. We stared around, as though that would help. We looked at each other. Dad's nose was red and dripping with rain. His anorak hood dripped onto his specs and into his eyes. The hand that gripped the plastic case in which Wainwright reposed, smug and snug ('"Pleasing views of the valleys." Hah!'), was dripping up his sleeve. 'I think we've lost the way,' he grunted. I had a feeble desire to shout 'Help!' It was just as well for the maintenance of 'face' that I didn't, because at that moment the three Greyhounds came pelting out of the mist.

Where could they have been? Why weren't they halfway to Garrigill by now? I didn't care about the whys and wherefores. 'Where are you headed?' I quavered. 'Over Cross Fell,' said the leader crisply over his shoulder. 'Come on, Jimmy,' he exhorted the tailing Greyhound, and they strode away and disappeared. Oh, God, no. Then Jimmy's voice came floating back. 'Come on, lads, if you're coming with us.' Oh, God, yes, please, thank you.

It's two miles from Great Dun Fell to Cross Fell, says Wainwright. We must have done it in twenty minutes. We slipstreamed behind the Greyhounds, stamping through bogs, skittering over rocks. Sheep went pelting away, the rain bounced down. Nobody said a word. All we had to do was keep Jimmy in sight, and that was absolutely as much as we could manage. At the stone shelter on the summit of Cross Fell the Greyhounds allowed themselves two minutes for a slug of water and a handful of raisins apiece. 'Coming on, lads?' enquired Jimmy. 'No, thanks,' our two voices panted as one, and all the Greyhounds laughed, the leader rather wolfishly, I thought. 'Path's there,' he said, pointing out a grassy descent. 'You'll meet the Corpse Road in ten minutes. Right's your direction then. Seven or eight mile to the George & Dragon at Garrigill. Should be there for opening time. G'luck,' and he cantered off tirelessly at the head of his little band.

The Corpse Road was used in olden times for carrying dead folk from Garrigill across Cross Fell and down to the Eden Valley, a ten-mile journey to burial in the nearest consecrated ground at Kirkland. Stepping out between mine shafts and levels along its well-founded course, we felt like two brands plucked from the burning. I don't suppose the Greyhounds thought for one moment that it was possible to lose the way on such an obvious path, but somehow we did. Rotherhope Fell, I hate you yet. It turned out to be a twenty-mile day. Splashing, squishing and footsore, cursing Wainwright and the mist, Dad and I crawled into the George & Dragon at Garrigill (one of the *Good Beer Guide*'s better suggestions) as night fell, as wet, tired and cross as could be. But better friends, somehow, because of it.

As St Cuthbert's Way approaches Teviotdale, the path declines to a boggy slide of cresses, black mud seeps and mountain-bike tyre tracks. Around Harestanes, Dere Street creeps off and disappears while I'm not paying attention. I blunder among rhododendron bushes in the grounds of Monteviot House, and break out into fields where a sudden afternoon hatch of tortoiseshell butterflies has filled the feathery grass. Tortoiseshells were the signature butterflies of childhood at Hoefield House. I would find them flicking their black and brown underwings behind the shutters, or hanging crisp and dry in the folds of a curtain in winter, or done up in a fantastically shaped cocoon.

I cross the broad Teviot on a wooden suspension bridge that swings and creaks with every step. On the far bank St Cuthbert's Way stops flirting with the south and heads decisively east over the curve of the earth towards Holy Island some fifty miles away. Broad parkland meadows fringe the river. It's a pastoral dream. In the shade of huge old specimen oaks graze flocks of fat white sheep, making me think of Cuthbert the young shepherd, who, according to his biographer the Venerable Bede, would rather pray than fool around with the other youths. What a goodie-two-shoes, what a milksop. But it wasn't always like that. Bede draws Cuthbert in early life as having been very much one of the lads, rather a cocky young shaver in fact, taking delight in 'mirth and clamour' and boasting of his prowess in running, jumping and wrestling. Then one day, something strange happened to young Cuthbert:

There were some customary games going on in a field, and a large number of boys were got together, amongst whom was Cuthbert, and in the excitement of boyish whims, several of them began to bend their bodies into various unnatural forms. On a sudden, one of them, apparently about three years old, runs up to Cuthbert, and in a firm tone exhorts him not to indulge in idle play and follies, but to cultivate the powers of his mind, as well as those of his body. When Cuthbert made light of his advice, the boy fell to the ground, and shed tears bitterly. The rest run up to console him, but he persists in weeping. They ask him why he burst out crying so unexpectedly. At length he made answer, and turning to Cuthbert, who was trying to comfort him, 'Why,' said he, 'do you, holy Cuthbert, priest and prelate! give yourself up to these things which are so opposite to your nature and rank? It does not become you to be playing among children, when the Lord has appointed you to be a teacher of virtue even to those who are older than yourself.' Cuthbert, being a boy of a good disposition, heard these words with evident attention, and pacifying the crying child with affectionate caresses, immediately abandoned his vain sports, and returning home, began from that moment to exhibit an unusual decision both of

mind and character, as if the same Spirit which
had spoken outwardly to him by the mouth of the
boy, were now beginning to exert its influence
inwardly in his heart.

This ticking-off by a three-year-old child, improbable
as it seems, changed Cuthbert's course entirely. He became
'devoted to the Lord', began seeing angels, and was able to
save several boatloads of monks from shipwreck and
death off Tynemouth by praying for the storm that had
scattered them to disappear. He did this in front of a mob
of rude fellows 'with angry minds and angry mouths'
who were jeering the storm victims from the clifftops.
Bede reports that Cuthbert's actions left the rustic rough-
necks 'blushing for their infidelity' – rather an agreeable
image. Not long afterwards, the prayerful teenager
decided to enter a monastery. On his journey he took shelter for the night in an abandoned shepherd's hut. Half
starved from fasting, the boy was singing a psalm to himself when he saw his horse reach up its mouth and tug
something from the straw thatch. Investigation revealed it
to be a nice hot loaf of bread and some meat, neatly
wrapped in a linen cloth. Who but God could have tucked
it into the thatch? Cuthbert shared the heaven-sent bounty
fifty-fifty with his horse, and next day rode on towards
Melrose and the monastic life.

In the soft parkland by the Teviot I stop to watch a ewe
summoning her daughter, almost as bulky as herself,
with a bleat as phlegmy and harsh as a sixty-a-day

smoker's cough. Then I stir my stumps and move on towards my night stop at Morebattle, still some ten miles distant at the feet of the Cheviot Hills. St Cuthbert's Way smuggles me through the landscape in a never-ending tunnel of trees, with occasional gaps inserted like windows to allow a glance out across the shallow bowl of Teviotdale. Through the afternoon I walk like an automaton, mile after mile, in that long-distance rhythm that scarcely acknowledges the ground underfoot. At last the path eases itself out of the trees, and I see that the rounded humps of the Cheviots have suddenly jumped up close, only a field or two away.

On a low ridge of ground between the Way and the hills, as plain as a toad, squats the graceless lump of Cessford Castle. This massive fortified house is the embodiment in blunt red stone of the roughness and wildness of the medieval borderlands. It rises nearly a hundred feet tall, dominating the neat grey roofs of Cessford village. The few windows it possesses are safely lodged in the upper storeys. Lower down are arrow slits. The sandstone blocks of which Cessford Castle is built are the colour of dried blood. Plenty of that precious commodity has been splashed around these walls, for sure.

Cessford is a fortalice, literally a small fort, a more imposing version of the bastles or fortified farmhouses one still sees dotted around the Scottish Borders. It was built around 1450 by the Ker family, one of the reiving or 'riding' families of cattle raiders and internecine thieves who made the Border too hot for the writ of any rule of law. The idea behind a fortalice was simple – to keep an enemy out, and if you have captured one, to keep him in. Hence the layout: a main stronghold keep up to six storeys high, with the two lowest levels barrel-vaulted for strength. At the bottom of the structure, a pit prison – a dark chamber, doorless and windowless, with a single tiny flue for ventilation and a hatchway in the ceiling communicating with the guardroom directly above. As the information board succinctly puts it: 'One way in and no way out.' Above the pit prison and guardroom, the kitchen; above that, two or three storeys of better-appointed rooms where the family lived. Outside, a gatehouse enclosed in an L-shaped wing, and a barmkin or defensive wall shutting the whole place off from the outside world.

Even on a lovely July afternoon such as this, the ruin of Cessford Castle could not look more cold and cheerless. It's hard to imagine spending a northern winter there. With logs blazing in the fireplaces, thickly woven hangings on the walls to keep the draughts at bay and plenty of looted food and wine, no doubt the Kers and their extended family and retainers (the fortalice could accommodate up to sixty of them) had a fine old time of it. As for the wretched captives down in the pit prison, left to cough or shiver or starve to death in darkness and their own excreta – it's better not to imagine their sufferings.

When they weren't fighting, robbing or murdering their neighbours, the Kers would sometimes raise their game from local to national level. A siege of Cessford by the English in 1523 has entered local folklore. The besiegers, commanded by the Earl of Surrey, managed to surmount the surrounding earthworks and the barmkin wall with scaling ladders. They got into the courtyard, but they could not break their way into the keep where the Kers had ensconced themselves, having blocked up all the windows with stones. The English brought up a couple of cannon and bombarded the lower windows until they had broken through one of the blockades. Some bold gunners then rushed forward with four barrels of gunpowder, which they managed to force through the rubble into the chamber within. But before the gunners could lay their fuses and fire them, the Kers with incredible coolness stove in the barrels and set fire to the scattered powder, which flashed up and burned harmlessly. The attackers at length acknowledged the stalemate, negotiations took place, and the Kers were permitted to surrender and leave the castle under safe conduct.

Cocking a snook at the Auld Enemy was one thing, but it was feuding with the neighbours that really floated the Ker boat. The riding families of the Borders were many and ferocious. As well as the Kers, there were Grahams and Charltons, Elliots and Robsons, Armstrongs and Burns, names you'll still find thick on the ground in these regions. The Scotts of Buccleuch were the special enemies of the Kers of Cessford. A blood feud between the two families simmered and spat throughout the sixteenth century. In 1526 one of the Kers was killed by a Scott retainer during a battle near Melrose. In 1548 the Kers caught Elizabeth Scott (born a Ker at Cessford) in her tower of Catslack and burned her alive there. Four years later, Walter 'Wicked Wat' Scott of Buccleuch, the head of the family, was run through and stabbed to death in Edinburgh's High Street by men of the Kers. And so on and so forth until 1607, when the Kers – under pressure, like the other riding families, to cease their lawless activities by order of King James I – abandoned Cessford and went in search of somewhere less primitive, in hopes of living less brutally.

A rainy, spitting morning lies over Morebattle next day. I'm tired after yesterday's fifteen-mile stage, but I pull on waterproofs and get going around midday. Cloud runs like smoke along the heights of Grubbit Law and drifts across the low green lands back towards Teviotdale. Foxgloves, harebells and lady's bedstraw border the path. Every leaf carries a string of rain pearls. Slugs with skins as black and shiny as PVC luxuriate in the wet grass. A bull bellows in long, horn-like blares from the side of Harrow Law a mile away, a trick of the air bringing the sound of its hoarse intakes of breath clearly to me. The Cheviot landscape unfurls ahead, steeply sloped, open to the sky, patched black with conifer plantations, humping and rolling like a grey inland sea. On the 1,207-foot summit of Wideopen Hill I register the fact that I'm at the highest part of St Cuthbert's Way, and also at its midway point between Melrose and Lindisfarne. But it's too rainy and cold to stop and stare round, or do anything other than trudge on. I slide down Crookedshaws Hill towards refuge in Town Yetholm, while the Cheviot-cross-Blackface ewes watch me from among the rocks, shaking the rain from their ears with a leathery flapping sound.

Next day I award myself a little breathing space. I'm rested and fed, and now I'm across the border in Northumberland the rain is holding off. Good friends come down from Edinburgh to walk a circuit of the delectable College Valley. In the evening we play a few tunes in the hotel at Wooler. Thinking of the double bosom of Yeavering Bell that we looked on today, and remembering a climb up there a couple of years ago, I ask Dave Richardson if he can play Alistair Anderson's haunting tune dedicated to, and named after, the hill. Dave doesn't have it at his fingertips. What he does have there is a bandbox-new Anglo concertina, one of the Rolls-Royce kind crafted by John Dipper, son of Wiltshire master-maker Colin Dipper. Dave has had it on order for a dozen years. He finally got his hands on it two days ago, and can scarcely bear to let go of it. As a nod to his native Northumberland he plays 'Proudlock's Hornpipe' and 'The Steamboat', and then an enchanting tune of his own, 'St Anne's Waltz'. The concertina has a tone unlike any I've heard before, rich and sweet, but subtle. Burr elm and Scottish elm, sycamore and mahogany have gone into its construction; nickel and brass and gold-tooled goatskin. It is an object that perfectly blends artistry and utility, but there's something more, a dignity about it, something generous and timeless. It will live on when Dave and I are long gone to dust. Somewhere in the world, in somebody's hands, this beautiful creation of an artist-craftsman will lift its voice, moving hearts and causing toes to tap.

Next day is another long one, but with journey's end in prospect tomorrow, the miles fly by. There is a pleasing juxtaposition near East Horton where St Cuthbert's Way crosses the Devil's Causeway, a Roman road that runs from Hadrian's Wall northwards through Northumberland to the mouth of the River Tweed. Saint spurns devil as the two roads diverge, the Way leading north-east towards the low and thickly wooded ridge of the Kyloe Hills. Here under a tremendous brow of sandstone that juts from the slopes of Cockenheugh I find St Cuthbert's Cave. A single slender pillar of rock joins floor and roof. The rock is scored with the graffiti of three centuries, its surface weathered into vertical grooves like ancient calcified skin. The cave might provide a shelter for a hermit of particular asceticism, I can see that. And one whom the birds were wont to feed, as in Cuthbert's case, could pass time here in prayer without starving. As to whether the saint actually repaired to this cave to watch and pray, only legend and local belief provide a narrative. Likewise with the tradition that monks from Lindisfarne, fleeing Viking raiders in AD 875, rested here with Cuthbert's body which they had carried away with them. No first-hand accounts remain. The Venerable Bede, who talked to those who had known Cuthbert and witnessed some of the miraculous happenings associated with the saint, had been dead for nearly a hundred and fifty years by that date. Suffice to say that the tradition lives on, passing from mouth to mouth.

Above St Cuthbert's Cave I cross a wide moor dotted with groups of cattle. Coming over the rise beyond I have my first glimpse of the coast and of journey's end, a long green neck of causeway running out to Holy Island. The tooth-shaped outcrop of Beblowe Crag rises from the flat pancake of the island, Lindisfarne Castle clamped to its summit like a tiny grey limpet. Through binoculars I can just make out the sandstone ruins of the abbey built, long after Cuthbert's time, on the site of the monastery where the saint practised his ministry. But no ordinary monastic life could hold a man with such an insatiable craving for solitude. First Cuthbert decamped to Hobthrush, a tiny blob of a tidal islet just offshore from the Holy Island monastery; then he withdrew to the harshest outpost he could find, the windswept, demon-haunted and utterly comfortless island of Inner Farne. I sweep the binoculars to the right and there it is, one among a crowd of low black dolerite shelves canted in the sea like a school of surfacing whales, baring white teeth of surf.

Inner Farne looks an impossible place for anyone to survive alone. But Cuthbert was incredibly tough, and incredibly determined. And he wasn't without assistants, though not in human form. Angels helped him build a stone shelter. An eagle caught a fish for him, and otters dried his feet with their fur after he had got wet while praying neck-deep in the sea. When he ticked off a parcel of crows that were stealing the thatch from the monastery guest-hut, one of the birds was so ashamed that it returned with a chunk of pig fat which Cuthbert used to waterproof his visitors' shoes (the saint was relaxed about receiving stolen goods, it seems). In particular Cuthbert loved the eiders that nested on Inner Farne, taming them and forbidding the locals to hunt them or take their eggs. Cuthbert's moral arm was long: the monks who followed his example and set up as hermits on Inner Farne after his death knew that to molest 'Cuddy's ducks' was to invite bad luck, a saintly curse from beyond the grave.

I watch the cars crossing the tarmac causeway, and try to distinguish the line of poles that marks out the pilgrim path across the sands. That muddy crossing is for tomorrow at low tide. I still have a few miles to go this afternoon, down to the rush and roar of the A1 and the Lindisfarne Inn where I can slide into a hot bath and get the aching legs up higher than my head. I've got to make some running repairs to my battered feet before I catch a ride tonight across the causeway to Holy Island and a church full of celestial music.

The last day dawns cool and overcast. I'm down on the shore by nine o'clock, and halfway across the sands by ten. It's easy going over the slippery grey mud, walking among uncountable thousands of worm casts and triangular gull prints. The barnacle-crusted poles that mark the way go by, one by one. I pass the rough wooden box on stilts into which unwary travellers caught by the tide can scramble. I pass the flowery dunes, the dark beach of the island and the village, and in gently freckling rain I reach the shore below the great ruin of the abbey.

The tide is on the ebb, but a narrow strip of water still separates Lindisfarne from Hobthrush Island. I'll just have to wait. I find a bench on the sward under the priory wall and sit looking down on Hobthrush. I count the chestnut heads and pale breasts of a huddle of wigeon on the island's tiny strand. A crowd of noisy black-headed gulls surrounds a solitary cormorant who stands stock still and upright like a black-clad policeman controlling a riot. An oystercatcher stalks the rock pools, its orange ice-pick of a bill lowered for action. Two common terns come planing in. They bounce to a landing, sweeping their wings up and back in an elegant vee before settling them with a shiver at their sides.

A grey seal swims seaward down the tidal channel, the movements of its powerful shoulders hidden under the water, its round bald head rotating as it stares at me with jet-black eyes. On the far side of the bay a pair of tall conical seamarks guards a sandspit, and as the tide recedes and more of its sand is exposed I hear the cooing and hooting of a colony of seals that has hauled out there. Now another spit rolls clear of the water, and immediately a little flight of Cuddy's ducks, chunky and dark with their heavy bills, comes down to land. The watery barrier has drained away now, and I can pick my path dry-shod out to Hobthrush.

The site of St Cuthbert's cell is indicated by a lump of masonry and a tall cross at the apex of the islet. I seat myself on the broken-down wall of the ancient cell. Did Cuthbert once kneel here, in a woollen shift if not in a cotton shirt? Did that intensely driven servant of the Lord ever allow himself to sit like this and think on being idle? I'm in a waking daze, torn between the cold reality of Cuthbert's rainy island and the cosy refuge of the skull cinema where a replay of last night's concert in Holy Island's church of St Mary is on permanent rotation.

If I could have picked one musician to crown my

journey from Melrose, it would have been Alistair Anderson, maestro of the concertina and Northumbrian smallpipes, composer and player of music inspired by a hundred different hilltops and landscape features of the Cheviots. And by serendipity and the luck of the pilgrim it happened to be Alistair himself, playing with a quartet from the Northern Sinfonia, who gave the concert yesterday evening in the little island church at journey's end. Not only that – 'On Cheviot Hills', Alistair's beautiful suite for strings and concertina, was the centrepiece, its heart and soul the air called 'Yeavering Bell', played with rare emotion by its composer.

I sit on the cell wall, content to wait till the rain drifts away, watching the preening of Cuddy's ducks and hearing the mournful hooting of the seals, playing back the highs and lows of St Cuthbert's Way to the inward soundtrack of 'Yeavering Bell'.

August

*The August man in thousands takes the
road to watch the sea and find the sun . . .*

ALONG THE SOUTHERN EDGE of the Lincolnshire coast, July tips into August. After weeks of cool damp weather I have found the sun, a hazy harvest one, and as I walk the sea wall along the eastern rim of England at low tide I watch the thin glittering line of the sea that must be – I check the map – five miles away beyond Old South and Blue Back and Gat Sand, the great mud and sand flats of the Wash estuary. It scarcely seems credible that at dusk in six hours' time the tide will have crept in across all those gently humped and gleaming flats, to mumble at the outer edge of the salt marsh and push its yellow froth fingers up the creeks as far as the sea wall where I'm walking.

Wherever the August man in his thousands has got to, he isn't here. He has taken the road to Skegness or the plane to Benidorm, and left this marginal land and sea to their great silence.

I set out an hour ago from Gedney Drove End. It's a functional place of brick box-houses and neat garden-centre plots, tucked in under the sea wall with no view of the sea. There has been a settlement of sorts here for the past two centuries or so, but a glance at the Ordnance Survey map tells you that Drove End lies in very strange territory. There are eight distinct places named 'Gedney' in this peripheral part of Fenland, and their names tell the unrolling story of how the land came into being. From

Gedney Hill, so-called – it is fifteen miles inland, but only about seven feet above sea level – the names, and the ruler-straight roads, march seaward in converging parallels by way of the freshwater swamp of Gedney Fen and the wide causeway of Gedney Broadgate, to Gedney village on the main road that links the ancient ports of King's Lynn and Boston. From Gedney towards the sea it is all land reclaimed from the sea since the mid-seventeenth century, the tale again being told in place-names: the drainage ditches of Gedney Dyke and Gedney Drain, the once-tidal stretch of Gedney Marsh, and finally a long, straight drove track to the end, the outer limits of dry land at Gedney Drove End.

I've been here before, thirty years ago, doing this self-same walk north-west along the sea wall. Very little seems to have changed in Drove End since then. The old New Inn has renamed itself the 'Wildfowler on the Wash', and is shabby and shut up tight. It's hard to keep a pub going nowadays in such a remote, cul-de-sac community. I remember Mrs Bills, the New Inn's no-nonsense landlady, and her stories of local goings-on. The best of them concerned the poaching prowess of her grandfather who would borrow his wife's shoes, four sizes too big for him, and wear them back to front so that the gamekeepers on the local estates couldn't follow his tracks. Poaching was always a major pastime hereabout, and it was the roguish tales of a famous poacher-turned-painter, Mackenzie Thorpe, that drew me to Gedney Drove End in the first place. I'd been given his biography, *Kenzie the Wild Goose Man* by Colin Willock, for my thirteenth birthday, and I was fascinated by its depiction of what seemed to be a kind of east coast Wild West populated with hard nuts and poachers, drinkers and scrappers, men digging fowling pits in the mud, film stars shooting on the marshes in pitch-black snowstorms. The 'Drove Enders' were a particular target for Kenzie's wit and mockery; he portrayed them as a wild bunch of back-of-beyonders, and in his day they probably were.

'Still are, some of 'em,' says the man in the paisley bandana who has leaned his bike in a gateway. He's sitting on the grassy sea wall, puffing a roll-up, a shovel at his side. 'Drove End born and bred, I am, so I ought to know! Slipper, they call me.' Slipper's face is deeply furrowed and weather-beaten. 'I'm just finishing this fag, and then I'm going to dig some old lugworms, see if there's any mackerel or bass about tomorrow. Fish for about three hours, and then you want to be walking back inshore. It's a long old walk with three or four bass to carry, and you don't want to be caught by that tide.'

Slipper gestures towards the distant mudflats. 'The secretary of the Wildfowlers' Association got caught in the tide out there a few years ago. Helicopter had to pick him up.' A cackle of smoky laughter. 'Well, he got so much stick over that, he stopped coming down here for a year or two.' Slipper sniffs me over, taking his time. 'Bristol? I was down there once about forty years ago, digging the West Docks as they called them. We built a roadway out into the Bristol Channel and worked from that. You had to clear three foot of mud before you got down to the red marl and the rock. When I saw all that mud I said, "Blimey,

I could've stayed in Drove End if I'd wanted to dig any more of that damn stuff!"'

Slipper flicks his fag end away. 'Don't go where I'm going,' he warns half humorously, ''cause I know what I'm doing, and you don't!' He trots down the outer face of the bank and on out across the salt marsh, moving rapidly with long strides among the sea purslane clumps and hopping across the creeks. I watch his active figure for a few moments, then move on. By the time I turn round and look for Slipper again he's a tiny stick-man a mile or more away, bent double and spading the mud for his lugworms at the very edge of the marsh.

Up ahead the sea wall runs straight to vanishing point. A pair of towers, one black, one white, indicates the site of Holbeach bombing range, a fat swathe of marshland five miles long, pricked out with lights and markers. Aircraft from European countries and the USA practise their bombing and shooting skills here alongside RAF planes. On my last visit a red flag had stopped me at the range boundary. A pair of USAF A10 tankbusters were making low passes, their cannon producing sparks on the ground and filling the sky with a noise like ripping canvas. I watched a band of pink-footed geese quietly feeding a couple of hundred yards away, apparently indifferent to the hellish racket around them. I'd been very struck by that. Something otherworldly about their coolness amongst the hardware and man-made thunder installed the pink-feet in a special niche in my affections. Today, a Saturday, the range lies silent. I walk on along the sea wall, thinking of the pink-footed geese that will be gearing up at this moment on their Icelandic breeding grounds for the long journey they'll be making in a month or so, bringing this year's youngsters with them to their wintering refuge in the Wash's vast food larder of a basin. I'll come back later in the year and look for them, I tell myself, and I feel a frisson of anticipation at the promise.

Contrasting worlds lie either side of the sea wall. On my right hand it is all dun-coloured salt marsh, sweeping away to the margin of the mud where Slipper is digging bait; then a wilderness of mud and sandbanks that seem to shift as I look across their heat shimmer to splinters of light at the horizon that are glancing off the all-but-invisible sea. To the left, twenty feet below the sea-wall path, the ground lies squared into enormous blocks of oil-seed rape, sugar beet, ripe corn and green potato shoots that stretch away past Dawsmere and Gedney Drove End, their rows and furrows arrowing inland until they merge with the flat skyline. Up till the mid-seventeenth century this was all salt marsh and floodland, unworkable ground where only fowlers and fishers could make a living. The Adventurers, speculators taking a punt on the reclama-tion of this watery wasteland, didn't invest in serious drainage until the 1650s, but then the Dutch engineering experts they hired began to claw back land from the sea and the flooded rivers incredibly quickly. By 1780 Drove End was a hamlet isolated at the outer tip of a salient of land reclaimed a century before, with the sea on three sides. A hundred years later, all the land around had been walled and dried out, desalinated and put under production, its Grade 1 silt feeding cattle on rich grass and growing the best cereals and vegetables in the country.

It's harvest time around the Wash. The farmers are hard at it, taking advantage of the break in the bad weather, and none of them will thank a nosy stranger for breaking in on their round-the-clock work. But I'd like to know what sort of people they are who stay here and work their socks off when everyone else has taken the road to watch the sea and find the sun. So I've fixed up a couple of farm visits in this strange Wash landscape with its submarine history and its uneasy relationship with the sea that may well claim it all back once more.

Tomorrow morning I'll be meeting George Hoyles and his son David at Monmouth Farm near Long Sutton, three miles away as the crow flies. Today I will just continue to walk this sea-wall tightrope, balancing between the salt marsh and the green and gold fields, until hunger and thirst reel me back to Drove End.

You reach Monmouth Farm down a long drove road flanked by telegraph poles. As with almost every farm in the region, the farmhouse is hidden away behind a massive dark hedge. These dense windbreaks, impenetrable to a passer-by, give an embattled air to the local farms in their level plain. 'When the wind's from the north,' says George Hoyles with studied understatement, 'it can get a bit draughty.'

We sit talking in the office at the entrance to the farm. An ancient wooden farm cart is parked outside, its boards beautifully bevelled, carved and painted, the elaborate curves of decorative lettering showing up through the dust: 'Hoyles, Monmouth Farm, Long Sutton'.

George Hoyles is seventy-two now, and has been farming here all his adult life. 'My first job on the farm?' He scratches his head and thinks back. 'Well, that would have been in the 1950s, leading the horses between the stooks in the harvest field. From that I graduated to the horse rake, then the plough. Those were Suffolk punches we used, or Percherons. Drilling with horses, or anything to do with managing them, was a skilled job. But once they were properly trained, they were wonderful to work with. A good well-trained horse could back a one-ton wagon into the shed on its own, without any human help.'

'What about now, Father?' asks George's son David, sitting across the table. David is forty years old, a thoroughly modern farmer, qualified to the hilt, keeping up to the minute with regular courses on how best to work Monmouth Farm.

'Oh, nowadays . . . It's getting a bit difficult for me on a modern tractor. Once it's going along, I can manage—'

'The problem your generation has,' interjects David, 'is turning it on! I mean, getting all the technology set up and going, the computer, the satellite systems and so on. But technology's not everything. People who buy land for investment and bring in professional contractors to work it – they don't know the land like we do. They don't know how to get the best out of it, what mistakes not to make.' He glances at George with a smile. 'I still rely on Father's advice. If you think of the seven-year rotation of the crops in a particular field, I in my working life have only seen one rotation. Father has seen seven or eight. He knows this land, every inch of every field. If I want to plough a field in this direction, or plant potatoes in that part of it, Father might say, "No, I wouldn't – that's too heavy, too strong, too high, it floods there," or whatever. You can't replace that sort of knowledge with technology.'

Monmouth Farm lies south of Gedney Drove End near the big farming village of Long Sutton, three miles from the sea. The land is several feet below sea level and dead flat. It was reclaimed early on, in the 1660s. The Hoyles family has been farming here for over a hundred years. In 1912, when they arrived, Monmouth Farm was 127 acres, but the family bought up land on their borders as it became free, and the farm is now about 550 acres of Wisbech Series Grade 1 silt, the best soil in the UK. In the 1960s they were buying land at about £350 per acre; nowadays it's worth fifty times that, thanks to outside investors with deep pockets pushing up the price.

As for what George and David Hoyles grow – that Grade 1 silt produces almost anything to a very high standard. 'People think of harvest as being in late summer,' David says. 'But ours starts in June and ends in December.' Peas, beetroot, wheat and potatoes are sold to the supermarkets, and mustard to the English Mustard Growers Co-operative. 'It used to be for Colmans,' says George. 'We'd take the mustard seed to old man Colman in Wisbech and he'd test the seeds himself, sniff them and taste them. It was personal. And the farmer's relationship with the buyers was personal, too. They knew the land like he did. They based their decision on the nearest railway station to the farm. When they heard "Hoyles of Long Sutton" or "Hoyles of Sutton Bridge" they'd want our business, because they knew the soil around Long Sutton railway station was Wisbech Series Grade 1, the best. But some chap down in the Fens on heavy land – they'd hear the name of his station, and they'd have no more to do with him.'

George admits that wildlife and the environment suffered from some of the chemicals they used back in the day. But his son doesn't have much time for those who make assumptions about the abuse of agrichemicals on today's Fenland farms. David does his courses, he consults expert agronomists. So do all the younger farmers he knows. He knows his stuff, and like most farmers he's pragmatic. 'You have got to target your pests,' he says, rapping emphatically on the office table. 'If there's an insect that'll eat another insect that does damage, why would I want to kill that "good" insect? It'll do my job for me, and save me money.'

Eco-friendly measures such as not ploughing right up to the hedge make sense in economic as well as wildlife terms. 'Even if you leave a nice wide headland round the edge, it takes up less than 10 per cent of a field. But if you plough it up you'll actually get 50 per cent less yield off that strip of land. Why? Because it's where you turn your machine each time you come to the end of the field, so you're compacting and crushing the earth and the crop won't grow properly in it.'

Monmouth Farm has marsh harriers and buzzards, and a good population of brown hares. 'Oh,' grimaces

David, 'we're overrun with hares. We had a hare shoot last February and got sixty-two. What we do get here is illegal hare coursing. It's very hard to prevent. You'll come across a group of men and their dogs in the middle of nowhere, and though you can't prove it, you know perfectly well that's what they're doing.'

Both father and son are saddened by what they see as the ever-widening gap in understanding between farmers and the public. 'Incomers to Long Sutton,' says George Hoyles, 'they tend to be from the cities, not country people. They complain about the church clock striking at night. They just don't seem to know what everyone used to know about the land, about what farmers do.' His son nods. 'Farmers are educated people these days. Not so long ago, schools round here would point the slower pupils towards farming. But it won't do now. There's no room any more for the traditional type of farm worker, a big strong man who's not too academic. Farms need people who are strong in computers, in mathematics – practical top-end people.'

The face of farming is changing all over the UK, and nowhere more than in these incredibly productive lands of the Lincolnshire/Norfolk border, reclaimed with such hard labour from the waste. Pick and shovel created them, sweat and muscle power brought forth their harvests. But that won't do today, not when you harvest your beetroot with a machine that costs a quarter of a million pounds and is chock-full of complicated technology.

'Technology,' muses David. 'Well, satellites bring me news of weather, of pests and disease outbreaks, right to my phone. The tractor computer programmes the seed drill to adapt the number and depth of the seeds to the varying soil quality across a field. We programme our machine wheels to follow certain lines so as not to compact the most productive soil where the crops will grow best. It's all incredibly detailed and technical. But so much of what we're mapping and programming is actually already there in the minds of the older generation.' He shakes his head ruefully. 'Technology is helping us to be busy fools, really, trying to understand all over again what our fathers already know, and what our grandfathers knew before them.'

I set out to walk the bounds of Monmouth Farm. In a shed beside the yard a group of silent, shy Polish men sit and stand in position around an elaborate machine, sorting beetroots that were harvested yesterday. In 1968, George has told me, Monmouth Farm employed sixty-eight people. Nowadays it's three, plus David Hoyles, and these half-dozen Poles who live for half the year on the farm, helping with the various harvests. Beyond the sorters a great wall of big wooden crates stands twenty feet tall. There must be three thousand crates altogether here. This is serious business, big business.

The clatter of the sorting machine fades as I follow the level lines of plough furrows away from the farm buildings. Glints among the remnants of pea stalks in the silty ground come from fragments of oyster shell, witness to the submarine origins of this heavy grey-brown soil. The farm's wind turbine emits a greasy whine as I walk past.

At the end of a track stands the £250,000 beetroot harvester, next to a tractor whose cab bulges with technological refinements. The seat alone costs £7,000 – 'You spend a lot of time in that!' – with its adjustable back and sides and its built-in heater. If you dial in your weight, the seat will reposition itself to accommodate your individual bulk. There are computer screens, harvester controls, keypads to adjust depth and rate of plough. Light and temperature controls. Radio and earphones. A camera to show the driver what's going on behind him. A StarFire 3000 satellite receiver. Perhaps, for all I know, a button for ejecting baddies through the roof. It is an incredible piece of space-age kit – and will be as obsolete as a Fordson Major in ten years' time.

I reach the boundaries of Monmouth Farm at Old Leam, a drainage ditch whose name suggests its antiquity. On the map it's a hesitant blue thread; on the ground, a ravine twenty-five feet deep and twice that across, its sides bright with wild parsnip, ragwort, poppies and thistles. Murky, military-coloured water slithers at the bottom. A dozen mallard get up with tremendous panic as my human shape looms over the rim. A reed bunting remains, preening its wing feathers on a reed stem, its pale breast winking as it flaps and fusses, giving out a chat-like click – *snip! snip! snip!* – every few seconds.

Old Leam meanders west through fields of crumbly soil. Each handful I pick up slips through my fingers like damp sand, sparkling with chips of mica and tiny shards of cockle and scallop shell. I follow the ditch as far as Lutton Gowts, the name signifying a 'go-out' or debouching of a watercourse. Here three streams once met the five-mile-wide estuary of the River Nene. Today the straightened and tidied-up Nene spans less than three hundred feet, lies three miles to the east, and flows due north into the distant sea through the best growing land in Britain.

I turn down the road towards Monmouth Farm and pass a home-made notice. 'Veg 4 Sale,' it says. 'Lettuce, toms, cues, cauliflower, carrots, onions, broccoli, leeks, potatoes, spring onions, apples, red cabbage, white cabbage . . .'

Later in the afternoon, back at Gedney Drove End, I climb up onto the sea wall and set off in the opposite direction to yesterday's walk. The green arm of the sea wall points ahead, crooking an elbow this way, then that, as it heads south-east for the mouth of the Nene. Two white pimples among the skyline trees gradually resolve into the East and West Lighthouses that guard the river mouth. I walk slowly, watching the fields of brassicas, peas and potatoes with newly opened eyes, picturing the high-tech ploughs and drills and harvesters that measure out their needs to the nearest millimetre, and the expert operators who work them in such cosseted isolation.

A herd of young cattle moves with great deliberation along the sea wall, grazing as they go, accompanied by a flock of a hundred and fifty starlings. Each time I come within spitting distance the birds all jump up together and fly off tightly bunched with a whirr of wings and a mighty twittering. They stick very close to the heifers, sometimes seeming to land momentarily on their backs. Are the cows exposing the hiding places of insects as they tug away the grass in mouthfuls, or is it the clouds of flies associated with the beasts that attract the starlings?

Beyond the salt marshes cloud shadows race across the incoming tide. Intermittent bursts of sun throw a spotlight on bright mauve swathes of sea lavender. When I reach the mouth of the Nene I find the banks gleaming with mud as thick and clotted as cottage cheese. The twin lighthouses stand one either side of the river. They were built in 1831, partly to shine their lights for sailors, and partly as seamarks to celebrate the opening of the Nene Outfall Cut or artificial mouth which tamed the sprawling estuary into a manageable channel.

When 24-year-old Peter Scott, son of the dead hero Scott of the Antarctic, came to live in the East Lighthouse in 1933 (rent: £5 per year), the tower stood right at the edge of the marsh. Scott was looking for solitude and the chance to indulge his two great passions – painting wildfowl, and shooting them – in one of the world's prime spots for both activities. But a piece of grim happenstance changed the course of his life. He was out wildfowling with friends one day when a solitary goose flew overhead. Everybody fired at the bird, and it crash-landed on a sandbank about five hundred yards away. They could all see it struggling with both its legs broken, its head raised as though looking for help; but it was out of shotgun range and the sand was too soft for anyone to get to it. Next morning Scott saw that the goose was still alive on the sandbank, keeping its head up. He was appalled at what he had done, taking himself bitterly to task for causing such suffering for his own amusement. 'I should not want this for a sworn enemy,' he wrote in his autobiography, *The Eye of the Wind*, 'and that goose was not my enemy when I shot at him – although I was his.'

Scott had already started a collection of wildfowl in ponds he had dug around the lighthouse. By 1939 and the start of the Second World War, he had turned his back on shooting and was concentrating on painting and preserving wildfowl and their habitats. He joined the Royal Naval Volunteer Reserve, serving in destroyers in the north Atlantic. After promotion to lieutenant-commander, Scott took command of a squadron of steam gun boats, fast armoured vessels that prowled the English Channel looking for trouble. They frequently found it. Scott's accounts of night actions with enemy forces in *The Eye of the Wind* are full of star shells, tracers, heavy fire and hair's-breadth escapes. Damage to the boats and their crews, sometimes fatal, was part of the job. Scott himself was famous for his coolness under fire. He was awarded a DSC and bar, and was three times Mentioned in Despatches.

After it was all over, the man of action went back to his paintbrushes and his wild birds with renewed determination to push forward the fledgling cause of conservation. Scott helped to found the International Union for the Conservation of Nature, and introduced the idea of compiling a Red Data Book of Endangered Species. He became in time the twentieth century's most influential champion of conservation.

The tiny, obscure village of Slimbridge lies on the

River Severn in south Gloucestershire, a few miles downstream of my childhood home at The Leigh. In 1946 Peter Scott selected Slimbridge as the birthplace for his particular baby, the Wildfowl & Wetlands Trust. Here he opened an observatory and some primitive hides, looking out on the reedbeds, flood fields and river landscape that I knew as a boy. It was a pioneering conservation effort, and Scott was soon presenting the BBC's first live wildlife series from his Slimbridge house. He became a worldwide star, but an accessible one. On family visits to Slimbridge we'd see him walking the paths, binoculars round his neck, polite and reserved, but responsive to questions about his beloved wildfowl.

'Please, Mr Scott, why are those ducks upside down?'

'Well, they're snipping off some weed to eat from the bottom of the pond. They'll be up in a minute, look!'

Scott cut a diffident and rather scholarly figure, modest in his manner. There was no whiff of derring-do about him. It would be hard to picture a less obvious war hero than this quiet painter and birdwatcher. But 'what I did in the war' was something no decent man cared to brag about back then. The war was still raw in many minds. Those who'd been through it knew all about it. Those who hadn't would never really understand.

It was the Old Peculier that eventually brought Dad into the open. The war, like GCHQ, was something he just didn't talk about, to his son or to anyone else – not his own part in it, anyway. If it hadn't been for Theakston's treacly, super-strong ale, he would probably never have unburdened himself. But two unaccustomed pints of Old Peculier snuck up on him one evening in August 1984, in the Buck Inn at Buckden after a hard day on the Dales Way. He'd been seven years retired by then, the war was forty years ago, and the heady beer finally opened the floodgates.

The Dales Way was our ninth and last long UK walk together. Each year of my increasingly frustrating tenure at the chalk face of teaching we'd done a hike in the long summer holidays – Pennine Way Part 2, the Two Moors Way across Dartmoor and Exmoor, a couple through wild Wales including the just-opened Glyndŵr's Way, the Cleveland Way round the rim of the North York Moors, and Alfred Wainwright's Coast-to-Coast. By 1984 Dad had had enough of the rainy moors and grey stone towns of the British Isles. If he was going to go walking, let it be where the sun shone and the local architecture piqued his interest and there were proper mountains to walk among. Somewhere away from here, over the sea where the colours were brighter. I agreed with that. So the Dales Way was to be our UK swansong, wandering up Wharfedale and Dentdale for a week in showers and sunbursts.

By now I'd more or less learned to bridle my overactive tongue when walking with my father. He could be abrupt when on the receiving end of naïve chatter. I discovered that it was better not to fill the silences with nervous babble or questions that Dad termed 'afas' – asking for

asking's sake. I didn't mind any more that he wasn't interested in pop music or modern poetry. He knew a whole
lot about a whole lot of subjects – politics, history, the Cold
War, the Moors, the Cathars, Gilbert and Sullivan,
Kennedy and Khrushchev. Now I felt at ease hearing him
hold forth, in his own time and at his own pace, after I'd
steered him into a conversational groove I was able to
follow. We were blokes, after all. We weren't going to dis-
cuss our emotions, or do the confessional thing. We were
still a long way from being able to embrace one another.
But we loved each other, sort of, in our British man-to-
man way. That was tacit. Meanwhile, let's talk. Another
pint, Dad?

'Lt-Cdr J. A. F. Somerville, Esq., R.N. (Rtd)'. Behind my
father's mysterious envelope titles, the shadow of the
unmentionable war stood vague and tall. The subject of
the Second World War was bound to fascinate little
schoolboys whose parents didn't acknowledge this great
clumsy elephant in the room. On Poppy Day our mums
and dads might refer to the First World War's slaughter on
the Somme, or the roses of Picardy. They might ironically
hum 'Tipperary' or 'Take Me Back To Dear Old Blighty'.
But they didn't talk to us about their own war. The same
went for their friends and acquaintances, all those
captains and corporals, wing commanders and second-
lieutenants. Our teachers, generally prepared to pass on
any number of facts about slugs and snails and puppy-
dogs' tails, grew reticent when asked what they'd done in
the war. All we had by way of reference were War Picture

Library comics, compact little 64-page Second World War epics set against real scenes of action – the Battle of Britain, Dunkirk and El Alamein, Russian and Atlantic convoys, Stalag Luft III and Japanese POW camps. Square-jawed sergeants, chirpy naval ratings and Brylcreemed rear-gunners on both Allied and Axis sides fought, wisecracked and died in the dog-eared pages of *Fix Bayonets*, *Call of the Sea*, *Fight – or Die!* and *Killer Squadron*. The black-and-white illustrations were hyperrealistic and accurate – unsurprisingly, considering that most of the artists had been serving under arms themselves only a few years before. From these grubby, much-thumbed little tenpenny shockers we learned about the war, about heroism and self-sacrifice on both sides of the conflict, about British pluck and stickability.

What my father and mother made of my letters from prep school, I can only imagine. I found it hard to fill two pages with news of my prowess ('Played cricket against Dumpton Grange, no wickets, no catches, out for 0 . . . did long division, but didn't get it . . . Osmond is a beast!?!'), so the lower half of the second side of letter paper was invariably graced with a scene of aerial combat from the Second World War inspired by the red-blooded action in War Picture Library. Spitfires vomited pecked lines of machine-gun bullets at earthbound Hun fighters belching flames in red and orange crayon. Speech bubbles floated from the cockpits of the combatants – 'Take THAT you NAZI RAT!!!' from the RAF pilot, 'AAARGH!?!' from his immolated foe. The Royal Navy never copped a bashing in these illustrations of mine, perhaps through consideration of my parents' feelings – Dad with his Royal Navy service in the Mediterranean, Mum in Africa and the Far East with the WRNS. The Army, of course, were beneath the notice of a war artist with Royal Navy roots such as myself. But the RAF seemed the very height of glamour and daring to me – though too brash and oikish for my largely naval family.

Few of us at that age had much knowledge of what it was that our parents had actually done during the war, still less of what the war had done to them. They wanted to put it out of their minds. They considered it bad form to talk about it, to 'be a bore about the war'. To us as children it was all about victory through righteousness. We didn't have an inkling of how the war had stayed bubbling under the post-war lives of our parents' generation until, much later, hesitantly and by modest degrees, some of them began to talk about it. And that happened with my father forty years after the event, sipping his second pint of Old Peculier in the Buck Inn and inching towards the elephant in the room.

In July 1936, three years before the outbreak of the Second World War, Dad was serving as an eighteen-year-old mid-shipman in the heavy cruiser *London* when she was dispatched to Barcelona, as part of an international maritime protection force that had come together to defend shipping and help refugees from the nascent Spanish Civil War. A *pronunciamiento* or uprising in the city by right-wing officers and men against the Popular Front government of Spain had failed, and the victorious

Republicans were executing large numbers of captured plotters.

'When we got to Barcelona there were hundreds of women on the quay, the wives of those officers, in a terrible state, all their children with them, everyone crying and grieving because their husbands had just been shot, or were being shot at that moment. One could actually hear the executions being carried out up the hill. The Republicans had set fire to a lot of churches in the city, and altogether it was really a hellish scene. We embarked the families and took them to Marseilles. I'll never forget the sound of those women, absolutely distraught.'

In autumn 1939, at the beginning of the war, Dad was in the Mediterranean as signals officer on HMS *Defender*, a D-class destroyer. In August 1940 he was promoted to lieutenant. By 1941 he was second-in-command of another destroyer, HMS *Hero*, escorting convoys to Malta and coming under heavy torpedo and dive-bombing attacks by German Fliegerkorps X. In April of that year a desperate situation unfolded in Greece, where an Allied expeditionary force was in chaotic withdrawal under German air attack after a short, disastrous campaign. The Royal Navy was assigned the task of evacuating as many soldiers as possible from the beaches and tiny ports of the Peloponnese peninsula – an extremely dangerous operation that entailed sailing from Alexandria after nightfall to the Peloponnese, embarking the exhausted and in many cases wounded troops, and returning to Alexandria before dawn. Any ship caught out in the open during daylight hours was likely to be sunk by bombers and torpedo planes.

On the nights of 28 and 29 April *Hero* crept into Kalamata at the southern end of the Peloponnese and picked up retreating Allied troops from the beaches in the ship's whalers, each night having to cut short the evacuation so as to be back in port before daybreak. 'At the time,' Dad said, 'I was waiting very anxiously for news of a great friend of mine, John Marshall. We'd been in Rodney term together at Dartmouth. We'd listen to classical music together, shared records and so on. He'd gone in to pick up soldiers from Nauplia in *Diamond*, another destroyer, two or three days earlier, and there hadn't been any news of them since then. We found out later that they'd been bombed and sunk. Some survivors were found in the water eventually, but not poor old John.'

At dawn on 20 May the Germans invaded Crete. On the night of the 22nd, *Hero* and another destroyer, *Decoy*, were sent under cover of darkness to the little beach at Agia Roumeli on the south-west coast of Crete to rescue the Greek king, George II, and his family. The royal party had escaped the closely pursuing Germans by crossing the White Mountains, no mean feat at the best of times. Dad vividly recalled the tension of the approach to the beach in strictly enforced silence, the drama of the enormous mountains silhouetted against the stars, and the relief as they got under way for Alexandria.

Two nights later *Hero* steamed into Suda Bay on Crete's north coast to land a detachment of commandos, part of a doomed attempt to shore up the rapidly collapsing Allied defences in the island. That was Dad's last sight of Crete for more than thirty years.

'Suda Bay was a vision of hell with burning ships and wrecks, thick oil slicks, floating bodies and black plumes of smoke going up on shore. I remember being turned out of my cabin in favour of a seasick commando, who was then sick in my canvas basin. I like to think it was Evelyn Waugh . . .

'In 1941 I owed my life, and in fact we all in *Hero* owed our lives, to our captain, one Hilary Worthington Biggs, D.S.O. He was the most superb ship handler under attack, when things got very nasty with explosions, wounded men, bullets zipping across the deck, lots of blood and so forth. Many times I was with Biggs on the bridge, watching him looking up at a dive-bomber and calculating exactly when and how to put the helm over so the bomb would miss us. The same when anyone saw a torpedo track heading for us.

'The one problem with Biggs was that he would not delegate. I was first lieutenant, but he couldn't bring himself to share responsibility for the ship with me or anyone else in those very dangerous situations. He had a bed of sorts made up on the bridge, and even had things rigged up so that he didn't have to leave his station when he had to answer a call of nature – he peed down a voice-pipe. Of course being on duty all the time made him very tired. And I found it very difficult, because Biggs wouldn't let me handle the ship under attack, and he didn't prepare me at all for taking command in an emergency. If he'd been hit or put out of action, I would have been responsible for the safety of the destroyer and the lives of a hundred and fifty men, and I just didn't know if I could manage that.'

In the event, Dad did not have to fill Cdr Biggs's shoes. The captain of *Hero* brought her through the crisis in Greece and Crete, safe though badly shaken and damaged by near misses. They went on to escort convoys into Tobruk and Malta – dirty, bloody and dangerous duty, but they survived it all. In 1942 Dad was back in the UK doing a long signals course, and he went on to serve in the Far East before the end of the war. But it was that hectic, dangerous time in the Mediterranean in 1941 that remained embedded most profoundly in his mind – the danger and tension, the awful sights and sounds, the importance of the task, the nagging doubts about himself. He buried those things deep, and I never heard a word about them until that beery night in Buckden.

Of the thirty-odd young men who had been in Rodney term at Dartmouth with my father, one-third died during the war, many of them as submariners. 'That's something one can't simply forget. It might seem odd, but the truth is that I've felt guilty about being one of the ones who survived, when so many of the best of us didn't. I've felt a need to justify my life, every day since then.' A long pause. 'Anyway . . .' He sighed, and began to talk of something else.

On the banks of the Nene I finish staring at Peter Scott's lighthouse and make inland for Sutton Bridge, yawning fit to beat the band. It's been a long day. A cloud of dust and a faint roar arise from the fields across the river. The high bar of the flood wall prevents me seeing the machine that's hard at work there, but I know it must be breaking the back of the oilseed rape harvest up at Lighthouse Farm. Tomorrow I'll be over that side of the Nene, walking the long miles of the sea wall round the edge of the Wash to the salty old port of King's Lynn.

Next day I sit in a farm office near Long Sutton, chatting to Julian Proctor and his son Stafford who farm the reclaimed land on the outer banks east of the Nene. Julian hands me a black-and-white photo. '1910,' he says, 'reclaiming the marsh as far north as Lighthouse Farm. That's how they did it back then.' Heavily moustachioed navvies are trundling barrows piled with sods of earth along narrow planks towards a new sea wall that rises in ragged segments. 'Three more sections have been reclaimed towards the sea since then: one in 1917, one in 1951, and lastly one in 1974 by our landlords, the Crown Estate. They built the new bank at the seaward edge of the salt marsh with material gathered by draglines from a "borrow pit" – that's a temporary scrape on the seaward side of the bank. A lot of mucky work!'

The section of marsh enclosed between the new and old walls remained a maze of creeks and mud for a year, only fit for grazing sheep. It's just one 370-acre piece in the complex jigsaw of fertile reclaimed land divided by successive sea walls which the Proctors have been working as Crown tenants since the 1960s. Now they farm about 2,250 acres, an enormous undertaking. About two-thirds is down to crops such as oilseed rape, wheat and peas that can be harvested with a combine; the rest is made up of root crops and a little grazing on the grassy sea banks. Back in the 1960s the Proctors employed up to twenty farm workers full time; today it's three or four.

I jump in Stafford Proctor's 4x4 and we go bucketing along the narrow roads to Nene Lodge Farm, a handsome house on the riverbank half a mile south of Lighthouse Farm. Behind the house stands an old red-brick store full of 1960s equipment – huge storage bins, vintage switch-gear and electrics, fifty-year-old ladders and rails – an antiquated set-up, but still well capable of handling the storage of the oilseed rape harvest. This is the moment during the farming year when the old store comes to life. The oilseed rape seeds, tiny, black and in immeasurable millions, arrive by tractor and trailer from the harvest field where they've already been separated from their parent plant by the combine's interior machinery. They drain like quicksand into a sump, the husks of cleavers and bits of leaf and straw forming white lines in the black sucking sea of seeds. Once drained into the store, eleva-tors carry the seeds upwards into the bins to await processing.

I follow Stafford up the railed stairs and along the walkways, looking down into the cavernous bins and their dark pools of seed. On the second floor is a Turner Dresser – a machine made in Ipswich in 1962, all of wood and tin. It stands solid and motionless as a rock until storeman Sid turns it on. Then it starts to shake, jiggle and thunder. Not so long ago the Proctors' entire corn harvest would have been fed through the Turner Dresser to shake the seeds free of the husks, separate out the chaff and reusable waste, and then grade the corn through a series of sieves. Nowadays the machine would struggle to handle even a quarter of the farm's volume of grain production. But it remains perfect for processing the mustard harvest, where the seeds all have to be meticulously sorted. So this rather homely looking and obsolete contraption is actually being used, by a twist of irony, for the most delicate work on the farm.

We drive across harvest fields that were tidal salt marsh a hundred years ago, towards the bank that the navvies were constructing in the old photograph of 1910. Over the far side stands another group of farm buildings. 'Kamarad Farm,' says Stafford. 'In 1917 there were masses of German prisoners-of-war camped at Sutton Bridge Docks, building that sea bank out there.' He points ahead to a long straight ridge at the far side of an enormous field. 'Every day they'd be marched the four miles to the works, each man carrying a brick in either hand. They'd add the bricks to a dump just here on the 1910 wall. And by the time they'd finished the new sea bank, there were enough bricks at the dump to build the farmyard here. Hence the German name of Kamarad ("Friend") Farm. In fact,' Stafford adds, 'there's a field nearby that's called "Verboten". I suppose the POWs must have taken a fancy to some crop in there and been warned off.'

Gaz the Yorkshireman ('My email name's yorkshire-pudding') is combining the oilseed rape in the wide flat field beyond Kamarad Farm. He's driving a huge CLAAS combine harvester, a machine making more of a hoarse whine than a roar. Slowly advancing, it trails a cloud of dust as its header reel whirls round and grabs up fifteen-foot lengths of tangled oilseed rape plants. Gaz brings the machine to a halt, and I climb on board beside him. It's surprisingly quiet in the cab. We are able to talk without raising our voices, and Gaz wears no ear defenders – and no mask, for that matter, though outside the air is thick with particles of oilseed rape and soil dust.

'So the knife cuts the crop and it's fed into the combine, as you saw just now. It goes up the elevator to be threshed; then the seeds are shaken out and separated, and the chaff blown aside by air blowers. I can set the sieves at different millimetre gaps from my seat here. The crop's then re-threshed to catch anything that didn't get caught the first time, and it can be threshed a third time if we want to. All the waste stuff – stalks and husks and so on – is ploughed back into the field behind the combine. Pretty much all recycled, which is what we want.'

We do a swathe and a half, and then Gaz lets me out. I watch as another worker brings up a tractor with a big open-topped trailer and positions it under the combine's flexible unloading auger. Gaz flicks a switch in his cab, and a solid black stream of rape seeds begins to pour into the trailer, thickening quickly and corkscrewing as it falls. In a few minutes the twisting stream slackens and hisses to a stop, and another load trundles off to swell the quicksand of seeds being sucked and elevated into the storage bins at New Lodge Farm.

Stafford drives me on up the east bank of the Nene, up over the 1951 bank beside Peter Scott's lighthouse, out to the grassy sea wall the Proctors built in 1974 with its footpath where walkers budge politely aside for us. A grassy foreshore leads out to olive-coloured salt marsh and then to the giant shores and skies of the Wash. Turbines whirl out at sea on the edge of sight to the north. Skegness is a faint white blur far up the Lincolnshire coast. The 272-foot tower of Boston Stump rises like a ship's mast from a grey ocean of flat land. In the east float the towers of King's Lynn, and far beyond them the red-and-white-striped cliffs of Hunstanton.

The sea seems so far from the land at this moment of low tide. But flooding is an ever-present threat, worsening with climate change. In December 2013 a sea surge almost overtopped the Nene's flood bank – it came within a foot or so of the top, and would have flooded Nene Lodge at the foot of the bank if it had spilled over. Stafford Proctor chairs the Wash Frontage group of farmers who are campaigning to raise the height and width of their flood banks to keep the sea at bay. At the opposite end of the argument are the conservationists, who urge 'managed realignment' by breaching the sea walls. This would inundate some of the farmers' reclaimed land, creating new grounds for wintering birds and allowing more salt marsh to form as a natural and effective buffer against the rising sea. These two positions are hard to reconcile. But then tensions between farming, food and the natural environment have always been at play in the Fens. Interesting times lie ahead.

*

The name of Peter Scott still carries huge clout in the conservation world. The sea-wall footpath that runs round the rim of the Wash from the lighthouse to King's Lynn has been named the Peter Scott Walk in his honour. I walk it in a dream. The stretch and weight of the huge skies gradually turning pink and green towards evening, the piping of plovers and oystercatchers from the giant mudflats and sands, dissolve time and space and a sense of my own significance. The miles pass under my boots, the air streams by. I chew samphire and scurvy grass and keep walking, the ascorbic savour hollowing my tongue and pinching away hunger. My feet ache, my eyes blur with wind tears, and yet the very last thing in the world I want is for this sea wall to reach anywhere. But it does, eventually.

I trail down the muddy bank of the Great Ouse to catch the last ferry of the day to the old Hanseatic port of Lynn. Green weed swirls along the timber baulks of the landing, and there's a sharp riverine stink of salt and mud. The little boat chugs me across the tideway. I scramble out onto the quay, still dazed with light and space, and head off down the narrow alley of Ferry Lane to collect my scattered thoughts and find a cup of tea.

September

September man is standing near
to saddle up and lead the year,
And Autumn is his bridle.

Down by the River Severn at Hayward's Rock, near Berkeley, on the first Monday in September. A narrow footpath enclosed in hedges full of black bryony shuttles me past the box shapes of the nuclear power station at the foot of Hamfield Lane. The path emerges onto an apron of coarse grass, beyond which the Severn is revealed at low ebb, a palette of ice-blue water, mauve sandbanks and terracotta rock. The river, at this point nearly two miles wide from bank to bank, is floored with jagged ledges of dark red sandstone and long, round-ended flats of muddy sand. The snub nose of Lydney Harbour lies opposite under the wooded ridge of the Forest of Dean, with dusky sandstone river cliffs a mile upstream.

I follow the Severn Way downriver. The grassy flood bank I'm walking is separated from the rocky ledges and sandflats of the riverbed, thirty feet below, by a thick shelf of mud and salt marsh where water left by the last high tide lies in long gleaming runnels. Suddenly there's a whirr of wings and a flash of halcyon blue and brilliantly polished copper as a kingfisher dashes the length of one of these indentations, then doubles back. It pitches beside the water in a patch of shadow, and immediately becomes invisible. I am really astonished that such a flamboyantly coloured bird can abolish itself in these surroundings of dull green and brown just by standing absolutely still.

How on earth does that happen? I look it up when I get home. It seems a kingfisher is in reality quite a drab little bird. When out of sunlight, its colours appear more like the leaves and twigs it spends its life among than the oriental jewel it resembles in flight on a sunny day. There is actually no blue pigment in a kingfisher's feathers, Google tells me. But the plumage does reflect shortwave blue light, which is scattered and thrown back by the microscopic particles that make up the feathers' structure. It's this blue light that one sees as the kingfisher's electric turquoise colour, instantly drawing the eye as the bird dashes by through a patch of sunlight.

Well . . . now I know. And it feels like a Garden of Eden moment. Science swallows magic. Will knowing what I now know tarnish for ever the glory of the drab brown kingfisher's beautiful flying suit of blue?

There's a persistent roar, low but strong, as the outgoing Severn is forced into a narrow channel between reefs. It seethes over a shallow sill of rock in a confusion of agitated wavelets. Typically of the Severn estuary at this state of the tide, one back eddy seems to be running inland against the seaward trend of the river – an optical illusion caused by the jabble of the water over rocks only just beneath the surface. I walk on, inhaling the well-known smell of the river – partly mud, partly rotting vegetation, partly the iodine savour of seaweed.

Severn House Farm lies tucked below the flood bank on the brink of the river where a long dark succession of sandstone shelves juts into the estuary. I stand above the farm, watching these ridges emerging as the water shrinks away. A wide pool begins to shape itself in the bed of the river, a depression slightly deeper than the surrounding rock plates and mudbanks which are now drying off. I notice a line of stakes running out from the riverbank, a row of black, seaweed-fattened stumps several hundred yards long. An old salmon weir, I tell myself, even before the memories come flooding in. 'Hayward's Rock', confirms the map, but it's surplus information. I have already recognized the place where I came on a summer's day twenty years ago to watch one of the last of the Severn trap fishermen go about his arcane business.

'Perfect conditions,' Deryck Huby had pronounced, squelching out through thick mud into the falling tide of the River Severn. 'Dry weather, no fresh water coming downstream, and a good north wind. Salmon swim into the wind, so they'll be over this side.'

This intimate knowledge of the ways of salmon had been ingrained into Deryck through thirty years of fishing the Severn estuary; not with rod and line but with hand-built weir, net and putcher. The salmon weirs, standing out into the water upstream of Bristol, had been a familiar sight to me since boyhood. Now, encased in thigh-high waders on a warm summer's afternoon in 1995, I was out in Hayward's Rock fishery, skidding on submerged rocks and probing cautiously with a stick in the wake of this master fisherman.

His weir, a zigzag fence of hazel poles and rods built earlier in the year, was festooned with bladderwrack and waste plastic, a pale forest of fluttering sanitary-towel liners, legacy of the last ebbtide. The weir enclosed a pool of water, leaving a narrow inlet near the bank for salmon to enter. At the outer edge, sixty putchers – trumpet-shaped cones, some woven of bamboo, others of willow – had been wedged in the gaps of the weir in a double rank to trap any fish that escaped Deryck's net. The falling tide tugged at my legs as I stood beside the weir, thinking of the Severn's last traditional putcher-maker, Dave Bennett, who had died earlier that spring. Some of these putchers could have been his handiwork. I'd spent a winter day watching him at his art, splitting and weaving willow rods, deftly shaping a piece of work in which simplicity was tied to subtlety, artistry to usefulness.

As the ebb gathered pace, Deryck splashed on out into the pool. Here he unshipped his home-made collapsible net and stood motionless up to his waist in the falling tide, watching the water with heron-like patience. He spotted the tell-tale bow wave of a salmon in the pool and flurried unsuccessfully after it with the net; then it was back to silent, endlessly vigilant waiting.

Salmon have been fished in this way for centuries as they travel from their Greenland feeding grounds to spawn in the headwaters of the Severn. Scooped up in the net or drowned head-down in a putcher, there should be no escape for any seaward-running fish at Hayward's Rock. But the art is almost dead. The wild salmon population has been decimated over recent years by wholesale trawling off Greenland. Farmed fish, on the other hand, are always available, and they are cheaper – if infinitely less tasty. The trouble and expense of renting a Severn fishery, building the weir and paying rates to the National Rivers Authority and the local council have made the short fishing season of ten weeks no longer worth the while of the handful of men who are still at it. And even if the finances did add up, the trap fisheries could not survive Environment Agency regulations that impose a quota of thirty salmon a season.

At the tide's lowest ebb Deryck moved out to check each putcher by hand. He had had three fish that week; but there were no salmon in the traps today. Deryck shrugged philosophically as we trudged back to shore empty-handed. 'It's in my blood, you see, this way of fishing. There's nothing like it, and never will be. I've loved every moment of these thirty years, and I don't want to stop just yet.'

When I made my way down to Hayward's Rock the following summer, I found the uprights of the weir still standing. But the putchers, and Deryck himself, were gone from the river. They'd never return.

It's the last Monday of September, up on Mendip, and the leaves are just beginning to turn. There's the faintest of yellow tinges to a few beech trees, some crimsoning to the leaves of cherry and dogwood. Jane and I are steering north-east tomorrow with autumn as our bridle on the long cross-country trek to Lincolnshire. Lindenshire, land of the lime trees, I've been reading; that might be the root of Lincolnshire's rather beautiful name. Actually, sober etymology says, it's more likely to be 'colony beside the linn, or pool'. But I like the notion of 'lime country'. We have an arc of woodland walks planned for the start of this season of the leaves – some ancient, some modern and some mythic. Lime country seems the place to start. We've had word of a National Nature Reserve, Bardney Limewoods, a skein of nine scattered woodlands out east of the city of Lincoln. Two in particular, Cocklode Wood and Great West Wood, have nurtured masses of small-leaved lime trees since medieval times, and maybe a lot longer. The small-leaved lime is just about the oldest established broadleaved tree species in Britain, and so far in our walking lives we've only seen them in ones and twos.

It's often tricky, walking in woods. In theory, Forestry Commission woodlands are open to all, but you never know how much monoculture you're going to find, how many corduroy battalions of unrelieved conifers blocking the views, or forest roads churned up by heavy machinery. The same goes for private woodlands. It's hard to find your way around a tangle of unmarked paths and rides when you can't relate where you are standing to any wider landscape because all those damn trees are in the way. And yet woodland walking has its own sharply defined charms. The weight and bulk of trees en masse, their dignity and grandeur, their movements that complement rather than mimic one another. The dreamy, seashore murmur of wind in the treetops. The sheer beauty of leaves by the million, the variety of their shapes, changing colour against the sky, or curled in gold and crimson along the paths to be shuffled through and kicked up by the inner (and outer) child. The birds that keep to the woodlands: the upside-down nuthatch, the scuttling treecreeper, the jay with its tearing and swearing. And above all, the feeling, never quite realized when walking in open country, of witnessing, as a guest might in one of those vast Edwardian country houses, the smooth working of an organism whose component members, above and below stairs, fit perfectly alongside one another with the efficiency of long integration.

Ladybirds are mountaineering up a tall hairy stalk of nipplewort beside a shady ride in Cocklode Wood. This late in the season there are only two flowers left to this plant, twin sprays of faded yellow petals at the end of one of the branching stems. One of the ladybirds has inched ahead of the others and is starting the long crawl out along the stem, hugging it tightly like an unconfident acrobat. It must be trying to get to the flowers, but why? Perhaps for a tasty snack of *Hyperomyzus lampsanae*, the aphid which favours the juices of the nipplewort. We look as carefully as we can along the spindly stems and tatty flowers, but there are no little green bugs to be seen. Maybe the ladybirds have a completely different agenda. Once again I am visited with a sense of rueful amusement that the greatest minds of unborn generations of humans are far more likely to find the cures for cancer, world poverty and religious intolerance than they are to discover what goes on in a ladybird's head.

Also along the rides are the fluffy wigs of meadowsweet, planting its pink feet as close to the wet ditches as it can get. Purple and white clover, knapweed like prickle-free thistles, St John's wort that can banish the blues. Some plants have shot up tall and are full of round green seeds – figwort, vervain, woundwort. Wild raspberries sprawl in low mats of foliage, the leaves sprouting from slender tendrils, prickly and red. Hundreds of orchid spikes remain, dry brown ghosts of the plants that flourished until a week or so ago wherever sunlight could get down to them.

And here, a few hundred yards into the wood, are the trees we have come to see. Their tall, straight trunks are grey and deeply fissured. They cast a cool shade and a calm aura in the wood. The heart-shaped leaves are darkly shiny, with no signs yet of colour change. They have finely toothed edges and a graceful pointed tip. They tessellate and whisper against the low grey sky, dangling on long pale stalks. No one can be sure how long small-leaved limes have been growing where Cocklode Wood and neighbouring Great West Wood now stand, but perhaps for seven or eight thousand years. They were among the first of the deciduous colonists, establishing themselves in well-drained soil rich in lime. They came to dominate the southern sector of Britain's native forest, romantically known as the wildwood, that covered these islands until man began to cut it down five thousand years ago. If, out of the blue, I had to name the archetypal tree of southern England's wildwood, I'd instinctively go for the oak. But the small-leaved lime – modest, unshowy, sombre – was queen of these islands for thousands of years before the monarchical oak made its play for supremacy.

There's more to the small-leaved lime than its antiquity. You can brew the blossom into a fortifying tea and twist the fibres of the bark into rope. Limewood is a top-drawer favourite with turners and carvers, being even-textured and of a beautiful colour between pale gold and silver. The Rotterdam-born master carver Grinling Gibbons used it for his solid and yet ethereally beautiful carvings of fruit, flowers and foliage. Three centuries after his knife shaped the limewood for works in St Paul's Cathedral, Windsor Castle, the Oxbridge colleges and elsewhere, it continues to glow with a subtle internal light, yellow-green and soft as oil, that echoes the muted light among the limes of Cocklode Wood.

Samuel Taylor Coleridge found inspiration in the lime woods. He told William Hazlitt that he sought out overgrown copses to walk in when he felt a poem was on the way, because the process of breaking a path through the tangled branches helped with composition. Maybe that was just naughty Coleridge teasing the impressionable young hero-worshipper to make his chum Wordsworth laugh. 'William, you'll never guess what I made him believe about me . . . !' But there's no doubt of Coleridge's appreciation of the beauty of a lime tree in 'This Lime-tree Bower my Prison':

> . . . *Nor in this bower,*
> *This little lime-tree bower, have I not mark'd*
> *Much that has sooth'd me. Pale beneath the blaze*

Hung the transparent foliage; and I watch'd
Some broad and sunny leaf, and lov'd to see
The shadow of the leaf and stem above
Dappling its sunshine!

Coleridge caught beautifully the movement of lime leaves on their long stems, though the foliage looks anything but transparent to us against this afternoon's heavy sky.

At the heart of the wood we find a beaten-up old army caravan silently rotting under a sallow. A stream trickles quietly through the trees, with the mossy trench of an ancient moat running away into stagnancy in neighbouring Great West Wood. In the fields beyond, a few brambly lumps and bumps are all that remains of Bullington Priory. Bullington was a rarity, a religious house founded in the twelfth century that followed the Gilbertine order instituted by St Gilbert of Sempringham. The Gilbertines were the only native British order, and they operated mixed religious houses for both men and women. It was quite a tricky undertaking to keep the sexes apart; the church had a wall built across it to divide the space between the nuns and canons of Bullington, but it was said that the canons were still subject to temptation when they heard the sisters singing. Three of the nuns were in charge of the priory's finances; the canons would pass whatever money they managed to collect through a little window to the unseen sisters, who squirrelled it away in a chest secured with three locks. Three separate locks, three keys in three chaste bosoms: Bullington's money was safe.

Bullington didn't possess enough land to maintain its prosperity, however. It spiralled into poverty after the Black Death depopulated the area, was soon classed as a 'poor nunnery', and was only a hollow shadow of itself by the time the last prioress, Mary Sutton, surrendered it to the Crown in 1538. Now it is nothing but hollows in the shade of the lime woods, allied to shadowy tales of a book of charms kept by the founder in a magical chest, and of unnatural luck that ran out on the priory.

We retrace our steps through tangles of undergrowth and swirls of make-believe. Where the decaying caravan lies at the hub of the two woods, we turn our backs on Great West Wood and retrace the rides of Cocklode, with the wind hissing in the Scots pines and stirring the dark foliage of the small-leaved limes across a greying evening sky.

There is something about an old dwelling in a wood. Something that trickles like cold water in the psyche. Today's children encounter the atavistic threat of the forest in the ultra-sinister fables of Jeremy de Quidt. But for me it began with the Brothers Grimm and their deep, dark forests. A brother and sister enter a forbidden hut. Hansel and Gretel approach the witchy house made of bread and sugar. Rapunzel pines in her doorless tower. The woods were full of sinister creatures: highwaymen and goblins, witches who caged and ate little boys, sly wolves who tricked young girls. Other writers, too, had their fingers on the quickened pulses of frightened little readers lost in forests of the imagination – Kenneth Grahame and his

Wild Wood, the ghastly spiders and man-eating trees of Tolkien's Middle Earth. Even stout-hearted, stout-booted Rupert Bear had his moments of terror at the hands of robbers and ravens deep in the trees. The woodcutter's axe and the footpad's wicked blade were never far away among those nightmare woods of childhood; also the malevolent glare of demonic eyes in the undergrowth. Saki's 'The Music on the Hill', an ominous fable of Pan-worship and 'the shadow of unseen things that seemed to lurk in the wooded combes and coppices', was a favourite if I wanted to give myself the creeps. The first time I read it, aged about twelve, I was on Bulbarrow Hill, lying all alone with a school-library copy on sunny turf beside a wood. I'd reached the denouement where the hunted stag, maddened by 'wild piping', lowers its antlers and charges through the whortle bushes at poor unbelieving Sylvia, and I can still taste the fright I got when at that moment a whistling sounded in the wood, and I looked round and saw a man standing there among the trees. I legged it so fast that I left the book behind on the grass, and had a tricky time explaining the loss away when I got back to school.

Every brave boy and girl with a head full of stories owes a debt to the dyers of Lindenshire; more specifically to those of the capital city of 'lime country'. Lincoln was one of England's chief cloth towns in the Middle Ages, and its dyers were celebrated for the secret recipe by which they blended the hues produced by the juices of two plants, woad (blue) and weld (yellow), to produce the famous Lincoln green. The dusky olive tint of Lincoln green was attractive and practical. And from the point of view of romantic young readers, a tunic and hose and a little peaked hat of Lincoln green were ideal garments for any outlaw wishing to live concealed in the shadows of Sherwood Forest.

Good old Robin Hood, the merry antidote to all those sinister echoes of the Grimm-style forest! Where he sprang from is a mystery. While fifteenth-century Merrie England was being ravaged by the Wars of the Roses, ballads of the exploits of the greenwood hero such as 'Robyn Hode and the Munke' were being sung in alehouses and halls up and down the land.

> Robyn toke out a too-hond sworde,
> That hangit down be his kne;
> Ther as the schereff and his men stode thyckust,
> Thedurwarde wolde he.

> Thryes thorowout them he ran then,
> For sothe as I yow sey,
> And woundyt mony a moder son,
> And twelue he slew that day.

> His sworde vpon the schireff hed
> Sertanly he brake in too;
> 'The smyth that the made,' seid Robyn,
> 'I pray to God wyrke hym woo!'

That was the stuff to give the people in tragic and uncertain times – plenty of swordplay, a bash on the head for the Sheriff of Nottingham, and a hair's-breadth rescue from prison by 'Litull John'. As for the sneaky monk who gave bold Robin away to the Sheriff, he didn't get the chance to bear his tales onward to the King once Little John and Much the Miller's Son had him in their hands:

> Be the golett of the hode
> John pulled the munke down;
> John was nothyng of hym agast,
> He lete hym falle on his crown.
>
> Litull John was sore agrevyd,
> And drew owt his swerde in hye;
> This munke saw he shulde be ded,
> Lowd mercy can he crye.
>
> 'He was my maister,' seid Litull John,
> 'That thou hase browyot in bale;
> Shalle thou neuer cum at our kyng,
> Ffor to telle hym tale.'
>
> John smote of the munkis hed,
> No longer wolde he dwell;
> So did Moch the litull page,
> Ffor ferd lest he wolde tell.

Hoorah! Bad cess to the tricksy monks and the bloody sheriff's men, and three cheers for the ale-quaffing, straight-shooting people's champion, Robyn Hode! This rumbustious ballad from olden time has a lot more blood and revengeful slaying in it than my childhood picture book *Adventures of Robin Hood*, but nothing has changed about the spirit of the stories, the constant action, the virility and good humour, and the inevitable come-uppance for anyone dishonest, pompous or treacherous. Robin lives, today as ever, not only in the films and on TV, in computer games and Hollywood merchandise, but in the imaginations of the boys and girls of this virtual and cyber century. Jane and I find that out when we get to Sherwood Forest for the next in our arc of woodland walks. Nottinghamshire kids run in and out of the visitor centre and in and out of the trees, and there is scarcely one of either sex who's not wearing a little peaked hat of Lincoln green or brandishing a sword or a shortbow. No one wants to be the nasty Sheriff. No one wants to be lanky Little John or fat Friar Tuck or moony Maid Marian. It is all Robin, Robin, Robin; Robin or nothing, Robin or bust. It's really astonishing, and it makes us grin.

Robin and his Merry Men probably had a hundred times as much forest to roam in as stands here today. Nowadays Sherwood Forest consists of about a thousand acres – two square miles of trees at most. Back in early medieval times the Forest was a royal hunting preserve of 100,000 acres, more than a hundred and fifty square miles of woods, waters, marshes, farmlands, grasslands and commons. Penalties for deer poaching had been softened or abolished under Magna Carta, so a Robin of the mid-thirteenth century would not have risked being hanged or blinded or having his hands cut off for that offence alone. An outlaw slippy enough to avoid capture and bold enough to hunt the King's deer could live well and in safety in Sherwood if he stayed on good terms with the locals – especially if he 'taxed' rich travellers in the Forest and shared the proceeds with the poor.

What else can Jane and I do but follow the Robin Hood Way? Within ten minutes it has led us to Sherwood's most famous living resident, the absolutely enormous Major Oak, which props its bulbous limbs and ancient hollow body on wooden crutches like a Peninsular veteran with a thirty-five-foot waist begging silently for alms. The leaves of its crown are turning gold. In one way it is the liveliest creature in the forest. If you could fit all the great tree's fissures and hollows together end to end they would measure many miles, and each inch hosts some species of wildlife, from burrowing wasps to spiders, hibernating beetles to ants and woodlice, nesting birds to lichens and algae.

Robin Hood held his court of thieves in the shade of the Major Oak some eight hundred years ago, stories say. But this tree could be twice that age. It is mightily gnarled and contorted, its bark stippled and wrinkled like rhinoceros hide. As we walk on along the broad forest ride of the Robin Hood Way we see more of these ancient oaks in the shadows on either side. Although they are the objects of everyone's admiration nowadays, they are the runts of the Sherwood litter. They only escaped the axe because the foresters of previous centuries deemed them too twisted or knotted or unshapely to be worthy of har-vesting. Some are limbless and barkless now, but today's forest workers spare them the chainsaw for the sake of the wildlife they shelter.

This central section of Sherwood Forest is known as Birklands in recognition of its silver birch trees, but it is all Oaklands for me today. While Jane walks the Robin Hood Way and looks for fungi, I leave the trail to move in closer to these arthritic old trees and feed my fascination with their humanoid forms. It's sloppy and unscientific to anthropomorphize a tree, but once you have begun to see faces in the trunks you are basically a hopeless case. Fly agaric fungi rise in the gold and green leaf litter under the trees, their sticky scarlet tops spattered with white spots, the remnants of the membrane that enclosed the young fungus. I think of *The Fairy Caravan*, Beatrix Potter's forgotten fable, and Paddy Pig seeing green things with red noses after eating 'tartlets' in the enchanted glades of Pringle Wood. You would certainly see a few funny things if you ate these pretty red and white toadstools, famous for their hallucinogenic qualities. But you don't need to risk madness and death by tripping out on fly agaric in the forest. A little dose of imagination will do the trick. Bent witches' noses, skinny arms akimbo, bulging eyes and howling mouths: as I stand and stare, they come out of the woodwork at me, literally. Here again are the frissons of 'The Music on the Hill', the glaring face in the wood, and the Green Man with his painful frown, his mouth stopped with greenery.

As with faces in the trees, once you begin looking for the Green Man in church roofs and porches, you can't help but find him. He is a wildwood presence, for sure. The masons of the Middle Ages slipped him in subversively among the prelates and princes they carved in churches and cathedrals up and down the land. But why? What was in the minds of those who fashioned his likeness, not just in the churches of medieval Britain, but on pillars and arches, doorways, bench ends and screens across Europe and further afield for thousands of years? They never said. No one wrote the Green Man's meaning down. No manuscript passed it on to us. It must have been so well known, so obvious, that nobody thought it worth their while to spell it out. So it remains, an enigma.

The Green Man takes many forms. The best known has a string of leaves growing out of either side of his mouth or his nostrils, more rarely his eyes or ears. But there are other versions: heads with features cleverly formed out of foliage, faces that peek out of garlands or roundels of leaves. The Green Man can be a woman, a cat, a dog or a lion. He vomits greenery, or is swallowed in it, or concealed by it. Or he himself can be part of it, his hair a burst of vine leaves and grapes, his beard curlicued with roots and tendrils of vegetation. At Sutton Benger in Wiltshire, the Green Man spews hawthorn. At Queen Camel in Somerset he is framed in vine leaves and clusters of grapes; in Tewkesbury Abbey he puts out oak leaves and acorns; in the Chapter House at Southwell Minster in Nottinghamshire he swirls with apple, iris and lace-edged buttercup leaves.

What is notable about the Green Man, whether he's shown as man, woman or beast, or something that is not quite any of these, is that he is rarely a jolly green giant. There is nothing serene or smiling about his face. It wears a grave expression. Often a deep frown draws the brows down. Sometimes there is a look of pain or astonishment, as if the vomiting forth of greenery is taking place against the Green Man's will. In the roof of Ely Cathedral he rolls his eyes and bares his teeth in a tongue-lolling leer. Or he can be a sly, fawn-like lurker and watcher among the leaves, a shadow of Pan among the Christian symbols.

Some theories have him as a lost soul in agony, the dark side of human nature that can be redeemed by being absorbed within the Church. Perhaps he symbolizes Christ, out of whose mouth issue the words that bring forth fruit an hundredfold. Certainly ideas of death and rebirth, decay giving rise to new life, seem implicit in this being who brings forth fresh leaves and fruit from his own body. Or maybe he is what he looks to be, a forest spirit smuggled into the church as a token of the ancient woodlands felled to make way for the monastic settlements and the earthy pagan religion that was cleared away along with the trees. That could explain his glazed, faraway stare, the animal forms he sometimes takes, and the compunction that the masons and woodcarvers so often felt to tuck him away in the shadows, where to recognize him is as shocking and compelling as glimpsing a dryad in a forest clearing.

I walk among the oaks of Sherwood, hearing again my father's growly voice at bedtime as he propped *Adventures of Robin Hood* on his brown corduroy knees. 'The Sheriff was seated at his dinner table, boasting that Robin Hood was afraid to show his face in Nottingham. Suddenly, through the window, flashed an arrow. It came to rest in the big fat goose the Sheriff was about to carve . . .'

Mmm, a big fat goose . . . I've never tasted a big fat goose. Daddy carves beef on Sundays. He's wearing his green and red tie, the one that makes my mouth taste of fish when I see it . . .

'The Sheriff turned white with fear. With trembling hands he unwrapped the message that was fixed to the arrow . . .'

Daddies don't turn white with fear. Their hands don't tremble. My daddy's hands are big and square. They feel a bit rough. His fingers are all spread out on the back of the book, but I can still see Robin's face peeking out among the trees.

' "It was I, Robin Hood, who won the golden arrow." '

Daddy's cufflinks are golden. If I shut one eye halfway, I can make them wink at me.

'Then the Sheriff's anger overflowed, for he realized that once again Robin had outwitted him.'

I'll be like Robin Hood. I'll be strong and brave. I'll outwit my enemies, tomorrow . . .

I didn't grow up to be like Robin Hood, though I had my outlaw phase. But by the time Dad and I finally abandoned the misty moors and cold rains of Britain and started to take our walks together overseas, we'd reached a better understanding of each other. I had swapped the frustrations of classroom teaching for the uncertainties of freelance writing, and he'd encouraged me all the way in a step that he must privately have thought over-optimistic, if not downright foolhardy. For my part I'd watched him struggle with the disgruntlements of retirement for the past fifteen years, sitting for brief periods of usefulness on this committee and that interview panel, chivvying the local council about holes in the road and dangerous junctions, dispensing good advice and practical help behind the scenes to friends and neighbours, sitting at his desk sighing like a weary guru as he composed letter after letter (typed, literally, letter by letter) to reluctant clergymen, urging them to promote the cause of Third World development among their flocks. The Rolls-Royce brain was in danger of racking itself to pieces on winding byways that led only to more frustration.

It was more of a release than ever to be able to put on the walking boots and, for one week at least, shed the bloody burden of responsibilities, petty or mighty, that he insisted on shouldering day in, day out. 'I've got to justify my existence,' was still Dad's day-to-day mantra, but in the sunshine on the hill trails of Spain or the vineyard paths of France, especially in the agreeable company of other like-minded people on the walking package holidays we signed up for, he could lay down that sack of woe too.

We got in five or six great expeditions before Dad turned eighty. There were June snowstorms over the Pyrenees as we guided ourselves round the gentian meadows of the Cerdagne region of south-east France. Another year, hiking the Corsican mountains, the guide got us lost the very first morning on a trackless hillside of burned and spiky maquis. All the walkers in the group emerged from that ordeal streaked with blood and charcoal like performance artists whose show has gone badly wrong. Corsica gave tough, sometimes risky walking for a 76-year-old; Dad found himself slipping and skidding on the steep limestone rubble of the gorge paths, and had to watch his footing every step of the way. But he found fascination in the gloomy village mausoleums lined out with black mortar, and humour in the unvarying solidity of the mountain menus – *soupe Corse* to start, monsieur; then an *omelette Corse* to follow, and to finish with, our spécialité nationale . . . *tarte Corse*.

We splashed out on a five-star trip to Provence the following year, and it turned out to be Dad's favourite, strolling at a civilized pace with civilized American walkers of a certain age through the vineyards and up the zigzag paths of the Vaucluse plateau. He allowed himself to relax into a week of sinful and expensive enjoyment, stopping for gourmet picnics under the fig trees, revelling in the knowledgeable conversation – stocks and shares, Republican nominees, the situation in the Gulf. One afternoon he spent walking alongside a courteous retired judge from Austin, Texas, exchanging experiences of the Second World War, something I'd never known him do with a stranger before. Best of all was the moment he had confirmation of his firm belief that all small French dogs are called Boby, all big ones César. In a vineyard near Roussillon a friendly little dog came bouncing up.

'Bonjour, Boby,' I greeted it. 'Ah!' cried its master in aston-ishment, 'vous le connaissez, monsieur?'

The expedition we took in Dad's eightieth year to the remote Alto Aragon region of northernmost Spain proved to be a step too far. The place was not the problem – Alto Aragon with its abandoned hilltop villages and lush flow-ery pastures was wild and remote and beautiful enough to soothe his cravings. His sense of humour remained undimmed, too. When I asked him to relax the grim expression he habitually assumed for a photograph, he rejoined, 'Reminds me of the instructions I once heard a gunner's mate issue to a funeral guard: Assoom a h'aspect cheerful, but subdood.' He could still crack a smile at his own expense, but things weren't right. The paths and the pace were just too much for him, the oldest in the group by thirty years. Uphill progress was at a snail's clip. 'Not enough steam left in the boilers, damn it.' His feet hurt. He felt uncertain of his balance, and that infuriated him. Years of neglect and rainstorm scouring had left the cob-bled shepherds' tracks in a parlous state, and I found myself walking as near Dad as I could get without irritat-ing him, one arm always at the ready to catch him if he should stumble. We began to lag behind the others, to be the tail-arse charlies, the ones that had to be waited for. It bugged Dad. He didn't want to spoil the party, and he didn't want any allowances made for him. At all.

At last he did fall, smack on his face on a tricky section of the GR1 below the deserted village of Pano. I wasn't quick enough to catch him. He bashed and bloodied his nose, gave himself a bruised cheek, and sat grumbling and apologizing while our concerned leader patched him up. 'Not really for the over-seventies, this sort of thing,' was his comment, sidestepping the fact that his next birthday would be his eightieth.

When we got home, Dad had some nice things to say. But that was the end of it. Those rusty boilers of his, as he saw it, were all out of steam.

We never went walking in company again.

From the ancient limes of Cocklode and the greenwood myths of Sherwood, Jane and I move south and west towards the last of our September wanderings in the woods. It's hard to imagine a nobler aspiration than the National Forest. A tidal wave of trees has been unleashed over the last twenty years to soften and repurpose a great scarred swathe of the industrial Midlands, two hundred square miles of western Leicestershire, southern Derbyshire and eastern Staffordshire, an area larger than the Sherwood Forest of Robin Hood's day. This was coal-mining and quarrying country. Now the coal mines are all but gone, the exhausted quarries gape in the hills, and many of the region's farmers are giving up, defeated by poor prices, thankless ground or a lack of interest on the part of the children who should succeed them. You can either leave the land like this, semi-derelict, worked out and burdened with the heaps and holes of outmoded industries. Or you can turn bare holes into lakes and black heaps into green hillocks. You can inspire local school-children and their parents to turn out and help plant the ravaged countryside over with ten million trees which will look beautiful, and bring in the insects and birds, and scrub the air clean of the global-warming gases carbon dioxide and carbon monoxide, and pump it full of oxygen instead.

Near the old coal town of Ashby-de-la-Zouch we walk the neatly mown rides of Willesley Wood, where a great hollow caused by the collapse of mine levels underground is now a broad lake among willows and cherry trees. Outside Ticknall the young aspens and silver birch of Windmill Wood flick their yellow-gold leaves in a sharp autumn wind as we pass by. Down in the south-east corner of the National Forest there are soft grass paths and hard cycle tracks to puzzle out in the coal country between Leicester and Ashby, where miners dug the black diamond seven hundred feet below the ground in Desford Colliery's seven seams. Two thousand men found employment there in the mine's heyday. A million tons of coal a year came out of Desford, but the miners' strike of 1984 killed it. We cross the open meadow where the pithead stood, and sit on a bench looking out across a subsidence lake to the spokes of the old pit wheel marooned on an island as a monument to the ghost colliery whose flooded levels lie below.

Our last afternoon is a blustery one. We set out from the village of Rosliston, in a tongue-shaped salient of south Derbyshire that pokes between Leicestershire and Staffordshire, to walk a skein of new woods scattered like jigsaw pieces across former farmland. Crab-apple leaves hang blackening, the fruits lime green and as hard as raw turnips. Elderberries are purpling up nicely, though still astringent in the mouth when tasted straight off the twig. Field maples and hawthorns redden in the hedges. In Long Close Wood the trees, planted eighteen years ago, are twice head high – cherries, silver birch, rowan with scarlet berries, guelder rose with crimson. The keys or winged fruit of the ash trees hang in pendulous lemon-yellow clusters, ready for a good blast of wind to whirl them far away from the parent tree and save humans the trouble of planting them.

On the way back we're cold enough to pull on gloves and woolly hats. But the grasses around Penguin Wood are bright with flowers that have not yet taken their hint from the weather about the changing season. The yellow froth of lady's bedstraw, the open pink mouths of campion and the fluffy heads of meadowsweet protest that it's summer still. The dull cold wind and a hint of low mist around the red and yellow woods proclaim the autumn. But Jane notes how starlings and crows are already beginning to flock together in the open fields beyond the woods, and she says it's a foreteller of winter.

Home from the woods. On the last day of September I head over the southern escarpment of Mendip and down to where the cider apples are ripe for harvesting. Here is an object lesson in the exercise of restraint. It's possible to enjoy the fruits of the cider-maker's labour as soon as he's finished his work. But it's better to wait, to possess your soul and your tastebuds in patience for as much as a twelvemonth to see the cycle round and the craft complete.

It is a slow, drowsy autumn afternoon on the Somerset Levels, and nowhere is slower or drowsier than Roger Wilkins's cider shed. The Wilkins family has been making and selling good cider at Land's End Farm in the hamlet of Mudgley, perched among its orchards a little above the flatlands of the Levels, since time out of mind. There is something peculiarly restful about this dark, creaky shed where a handful of men sit well back in plastic garden chairs, ruminating and sipping. The air is full of the heady, sweetly rotten smell of cider, a drink that does for its component element what wine does for grapes or poteen for potatoes, transforming humble fruit into liquid magic. In the small glasses over which the purchasers are humming and hawing, the cider has a milky hue, an absinthe shade with a faint greenish tint. Farm cider and absinthe have quite a lot in common, in fact: they both carry a great weight of myth, they are both very much a drink for specialists, not to say connoisseurs, and over-indulgence in either packs a weird psychological punch which has little to do with the well-known effects of alcohol.

There are two big dark oak barrels at one side of the shed, sweating slightly in the still air. They smell of saw-dust, a faint whiff of wood resin and a tarry tang of former ciders. One holds the sweet (not that the uninitiated palate would find it so), the other the dry. The trick is to mix the two – a trickle from this, a smidgeon from the other – until you have exactly the taste that suits you. That can vary. A few weeks on the sweet cider can sicken a devotee. Only a counter-blast of proper cheek-sucking dry will restore palate balance. Most purchasers are content to let Roger mix them a medium blend, enough dry to get the cidrous tang, enough sweet to ease it past the gullet.

Most of the point of an afternoon in the Wilkins cider shed is to soak up the timeless atmosphere. If anyone wants to know how the world outside is wagging there's usually a newspaper about, generally the local rag, and one of the customers will be hunched over that, glass in hand, stertorously breathing, perhaps reading, probably not. Mostly people chat from their chairs, or propped up by the crook of an arm against the barrels or a wooden beam – lazy talk, inconsequential and purely local, but sometimes digging down to the pips of Levels scandal. Everyone sips as the talk goes round. Cheeks get a little redder, laughter a little thicker. These are moderate consumers: a couple of local young men, a farmer friend of Roger Wilkins, a chap down from Bristol thirty miles away to buy a gallon for a summer party he's having in his city garden tonight. Roger himself could put away a couple of gallons a day when he was younger. He has slowed down a lot since then, and says cider has never done him any harm.

The Wilkins method of making cider is the old one, and the best. Roger grows and hand-picks his own apples, and buys locally what the Land's End trees can't produce. He looks for a good balance of bitter sharp and bitter sweet. The apples are crushed in a big hydraulic press to make pomace, traditionally a process done by piling up a 'cheese' with alternative layers of straw or sacking and apples. The apple juice runs down the outside of the 'cheese' in streamlets and trickles, to be collected, fermented and stored in the big barrels until the old oak, the apple juice and the slow concentration by sweating produces a drinkable, and then a delectable cider. The wrung-out pomace goes to the pigs and cows. Nothing's wasted here.

Some use the term 'scrumpy' to describe this kind of home-made cider. As a generally accepted term for farm-produced cider with no added sugars or colourings, that is fair enough. Others call it 'rough cider', a term that conjures up a palate-curdling brew with a vinegary savour and a kick like a mule. The kind of cider produced by Roger Wilkins and other small-scale West Country cider-makers is nothing like that at all. It is not exactly smooth, but it's very far from rough. It doesn't slip down without touching the sides; instead, there is a faintly but pleasantly abrasive effect along the gullet, and a sharp after-tang on the palate that dissolves almost instantly to release a burst of intense apple flavour. A glass of this cider is not clear, as a commercially produced sparkling cider is see-through clear, but its distinctive cloudiness is not of the 'nasty floating bits' type; it is more of a thick haze, of the type seen on autumn afternoons such as this one, hanging just above the water-filled ditches and damp meadows of the Somerset Levels – an indicator of lushness rather than of corruption. Hold a glass of Wilkins's cider up to the light and you will see a diffuse glow, as though looking up through deep water at the sun.

In times past the cider-maker might chuck a handful of raw beef – some say it was a skinned rat – into the vat to give the cider a bit of body. If Roger follows that tradition, he's not saying. One old custom he does keep up, though, is the wassail on a Saturday evening in January. If you don't honour the trees with a wassail, they won't produce for you. You drink, you dance, and you sing:

> *Old apple tree, we wassail thee,*
> *And hoping thou wilt bear*
> *For the Lord doth know where we shall be*
> *Till apples come another year.*
> *For to bloom well, and to bear well*
> *So merry let us be;*
> *Let every man take off his hat,*
> *And shout to thee, old apple tree!*
> *Old apple tree, we wassail thee,*
> *And hoping thou wilt bear*
> *Hatfuls, capfuls, three bushel bagfuls –*
> *And a little heap under the stairs!*
> *Hip! Hip! Huzzah!*

October

The man of new October takes the reins
and early frost is on his shoulder . . .

FIVE O'CLOCK OF A freezing October morning on the North Norfolk coast. In the RSPB car park at Shepherd's Port it's pitch black under a sky brilliant with stars. For the first time since that blasting Midsummer's Day on Foula I'm kitted out in woolly hat, scarf, thick gloves, thermals and overtrousers.

The narrow path winds southward into the dark of Snettisham Reserve. The light of my torch rims the bushes and grasses with a white coruscation of frost. To the west, over the giant square-sided basin of the Wash estuary where the ebb tide is halfway out, everything is black. In the eastern sky sails a planetary trio – Venus shining like a tiny, intense searchlight, Jupiter almost as bright at her right hand, Mars a little below. To the south Orion the Hunter straddles the sky segment towards which I'm walking. Every star that frames him is clearly pricked out. Six points of light to his left form the curve of his bent bow. Four more, high over his right shoulder, shape a pair of archer's fingers reaching for an arrow from the invisible quiver at his back. There are three silver studs to his belt, from which depend three more, suggesting what might be a dagger or a dangling penis. He is a virile figure, huge and dominant in this early winter sky.

It's absolutely silent. All of a sudden a tremendous uproar makes me jump and curse, bursting out on my right hand in a hysterical high honking and calling and a clatter like a paddle-steamer thrashing through water. A moment later the heavy beating sound transfers itself to the air and saws away to the south. I flick the torch across my map and the mystery is revealed. A string of flooded gravel pits, unseen by me in the dark, edges the path at this point, and a flock of big birds has been roosting there overnight, probably pink-footed geese not long arrived from Iceland and feeling jumpy. Either the modest light of my torch or the crunch of my boot in an icy puddle has set them off. The surface of the lagoon ripples and rocks for a few seconds in the torch beam, then the empty water settles and a few white feathers spin slowly to a standstill. I curse myself. I can hardly bear to think that the first contact I have had this year with my 'special' geese, the ones closest to my affections, has been so damned clumsy.

I switch off the torch and stand listening for any sounds out on the mudflats. So far this season about five thousand pink-footed geese have arrived to overwinter at Snettisham. The majority will have spent the night out on the flats, roosting in safety from foxes and men. The tide is still withdrawing from the land, inching its skirts further away with every ripple. Even at half-ebb, if those geese are roosting anywhere near the tideline they might be as much as two miles away from me. It's still early in East Anglia's wintering season, but by this time in late October there could be up to fifty thousand pink-feet already spread around the vast tidal basin of the Wash, with seventy thousand acres of mudflats to choose from. However many there are, and wherever they have gathered in the profound darkness out west, there's no sound from them.

Rotary Hide looms up eventually, a cold wooden box of a place ideally sited in the flood bank. One can sit on a hard wooden bench at the hinged windows looking east over the lagoons to trees and low hills, or cross the hide to position oneself at windows that look west into the flat empty wastes of the Wash. I wonder if Slipper is out there in the darkness with his nets or his 12-bore goose gun, beyond the sea wall at Gedney Drove End a dozen miles away. With the torch off and my eyes adjusted to the dark, I can see the harsh white lights of ships riding out the night along the western horizon. I turn my back on them and settle at the eastern windows, hugging my cold hands into my armpits and looking out at a sky where Mars has already vanished into a faint green haze just above the skyline.

From the southern corner of the gravel-pit lagoon in front of me comes the conversational gurgle of geese, probably the flock I disturbed half an hour ago. They are alert, croaking and creaking to one another, growing more excited and muddled as the light broadens and pales. Soon I begin to pick out individual voices – a squeaky soprano complaining, *eee-eee!*; a hoarse basso profundo suggesting a goose with a chesty cold, *uurgh, uurgh!*; the quavering falsetto hoot of a girl laughing her head off, *ooh, ooh, ooogh!*; the downbeat and plaintive honking of a tired man wishing everyone else would just shut up, *ee-aagh!* I'm well aware that geese are wild creatures and one shouldn't anthropomorphize them, but it's impossible to resist as they gabble and shuffle about in long black lines on the water. 'Oh, come on, let's get going.' 'No, love, give us five minutes more, eh?'

Some communal decision is taken, and with a great kerfuffle they all heave themselves into the air and disappear, seventy or eighty big, powerful birds. A couple of wigeon go whistling across the dawn sky, now pink and green with a brushing of gold on the undersides of the cloud streaks. A cormorant lifts off an islet in the gravel pit and flies low and direct across the flood bank towards the sea. But the pink-feet are not passing in front of me in those characteristic long straggling vees, heading inland to break their fast as I was expecting them to do. Where have they gone?

The eastern sky is lightening over the sugar-beet fields that lie a few miles inland. The beet harvest has just got under way on the farms around the Wash – I saw it myself yesterday, big yellow loaders and scarlet harvesters and long mounds of pale beet stacked ready for the lorries. That's why the pink-feet are here, primarily. Half the world's pink-foot population, up to two hundred thousand geese, come to winter in this particular corner of these islands because the fields are full of the nutritious tops and tails and slicings of sugar beet left behind by the harvesters. A cap on EU beet production introduced in 2005 saw the numbers of overwintering geese fall dramatically in the following years. The pink-feet found empty fields where they'd learned to expect a feast, and went elsewhere. Now, ten years later, the beet farmers have just learned that the cap is going to be lifted and they will be able once more to produce as much sugar beet as they like. Maybe next winter will be a bumper one for pink-footed geese around the Wash. That's all in the future – but meanwhile, why aren't the geese from the lagoon hot-winging it to breakfast in the beet fields?

Behind me a gabbling commotion breaks out, along with the sound of wings energetically flapping. I turn my head and see that the pink-feet from the lagoon have just come in to land on the mudflats to the west, about two hundred yards away. They have done the unexpected and headed seaward instead of landward – out to the safety of the glutinous mud desert, rather than inland to the rich beet pickings for which they must be starving after the long hungry night. They are standing erect and watchful, heads held high, flapping their wings over their barred grey backs before settling them at their sides. They look so beautiful in the cold light. This time I'll be more careful of their sensibilities. I cross the hide and gingerly ease open the hinged window that looks out on the Wash. In spite of my care the heavy frame clips to the wall with a soft *thunk*, and half the pink-feet turn their heads my way and squawk. Almost certainly they are freshly arrived and full of trepidation, hungry but wary, uncertain about the noises coming from the hide they'll have to pass over on their flight to the beet fields, unwilling to retreat all the way out to their massed brethren on the tideline.

A yawn splits my face in two. The sun lifts itself over the trees in the east, seeming to spin like a golden Catherine wheel. I've been on watch for a couple of hours now, and I'm stiff and cold. I shut the windows as quietly as I can and leave the hide. Now I'm in plain view of the pink-foot flock, and they don't like what they see. I head back north up the flood-bank path and the geese patter anxiously along parallel with the shore and slightly ahead of me, caught between the devil and the deep blue sea. I don't want to drive them, and they don't want to be driven. I want my breakfast of bacon and eggs at the Rose & Crown in Snettisham, and they want their breakfast of beet tops in the fields around the village. How are we going to accommodate each other?

The pink-feet solve the problem. The leading goose gives out an echoing cry and they all get airborne, beating along and gaining height. They wheel seaward, making a circuit perhaps half a mile in diameter before flying in across the sea wall in two loose vee formations, three or four hundred feet up and well beyond the Rotary Hide and its attendant demon. As I walk back past the gravel pits and across the common among wild rose bushes weighed down with fat scarlet hips, a long look seaward shows me a vast swirling ball of little birds high in the air – knot, probably, ten thousand or so of them – and the familiar wavering lines of hundreds of pink-footed geese above the mudflats, heading inland for the beet fields, another night survived.

It's ebb tide on the North Norfolk coast, and ebbing daylight too, the October afternoon breezy and cold under one of those faintly gleaming silver skies. I set off eastwards along the coast path from the quayside at Burnham Overy Staithe, thirty miles from Snettisham as the pink-feet fly. Black and orange cattle graze the freshwater marshes on my right hand. Invisible beyond the salt marshes and sandflats to the north, the sea is sliding away from the land. The River Burn follows it out down the gliddery grey windings of Overy Creek, and I follow the river. Along the reedbeds a million feathery heads toss and whisper together. The outgoing water smells of mud and salt. Slow-moving and heavily charged with silt, it has a long way still to flow, well over two miles of turning and twisting between salt marsh and sandbanks, before it reaches the sea between Gun Hill and the dunes of Norton Hill. It runs with a smooth power, tilting the boats onto the mudbanks of the creek with a rattle and chink of halyards against hollow metal masts.

A land that is thirstier than ruin;
A sea that is hungrier than death;
Heaped hills that a tree never grew in;
Wide sands where the wave draws breath.

Algernon Swinburne is unfashionable these days, but someone likes him enough to have posted a verse from 'By the North Sea' on the quay. The waterside buildings of Burnham Overy Staithe are solid sheds of flint cobbles and tarred brick. The staithe or wharf faces out towards the sea, source of its former prosperity; but it was killed off as a viable trading point in the 1850s when the railways sneaked up from landward and stabbed it in the back.

The Tudors reigned in England the last time a ship

passed this place with a cargo to unload at the wharves of Burnham Overy Town, a mile inland from Burnham Overy Staithe. Once upon a time ships could navigate the River Burn all the way up to Burnham Thorpe, birthplace of Horatio Nelson. England's great hero had his first childhood sailing adventures on the River Burn and Overy Creek. But today his native village lies five miles from the sea. Ten settlements in a tight little ring hereabouts share the Saxon title 'Burnham', meaning 'village on the river' – Burnham Staithe and Burnham Town, Burnham Market and Burnham Thorpe, Burnham Norton and Burnham Sutton, Burnham Overy and Burnham Deepdale, Burnham Westgate and Burnham Ulph. The sea has deserted them all. And it's not just the Burnhams, by any means. Wells-next-the-Sea, Stiffkey and Morston, Blakeney and Cley next the Sea, Salthouse and Weybourne – they lie like beads on a string along the coast, a line of red-roofed, flint-walled villages whose old wharves and warehouses face the sea across a mile or more of salt marshes, 'next-the-sea' no more.

On the banks of Overy Creek black-headed gulls in white winter hoods bend at the waist in their anger and scream at each other like fishwives. The seabank path is slippery with grey sticky mud. I ought to be watching my step instead of trying to scribble notes as I walk. Twenty or thirty pink-feet materialize above the quay and come swinging across the salt marshes towards me, the first flight of the afternoon as the geese begin to leave the beet fields and head for the coast and their night's roost. I lift my binoculars to the sky to admire the big birds, my feet slide from under me, and I fall flat on my back with a thump that drives the breath out of my body. The pink-feet hoot and cackle as they pass overhead. I curse my inattention, scramble up, retrieve my notebook and pen, and trudge on down Overy Creek. By the time I reach the long ridge of sandhills where the sea lies, the sky is streaked with vees of geese. The quayside buildings of Burnham Overy Staithe are tiny rectangles now, crouching at the landward edge of a solid mile of dun-brown salt marshes.

Longshore drift is the villain of the piece. The coast around Cromer, thirty miles to the east, is composed of crumbling cliffs packed with pebbles of flint and stone. North Norfolk's tides push westwards, and as they do so they carry an enormous weight of flint pebbles from those cliffs along the coastline before dumping them in long reefs piled up underwater, a little way offshore. The rivers coming down from the hinterland, Burn, Stiffkey and Glaven, are thick with silt, but they can't discharge it into the open sea because the pebble reefs block their passage. So they dump the silt at their mouths, a mineral-rich carpet of nutrients just perfect for the growing of salt-marsh plants like sea purslane and sea lavender. The salt marshes thrive and spread, the rivers are forced to cut ever narrower and more tortuous paths through the vegetation. The waterways silt up and shrink, and the little settlements at the former river mouths grow ever more distant from the sea that was their lifeblood. Lucky for them that they are so fetchingly pretty, and that London is only a few hours down the road. Tourists and second-home owners have replaced the shipbuilders and merchants, the boat owners and the deep-sea fishermen who worked for them. Apart from the winter-long absence of lights at the flint cottage windows, the villages have a sheen of prosperity these days. Builders and refurbishers are doing very nicely, thank you. Burnham and Morston, Brancaster and Cley are far neater and tidier now than ever they were when the sea lapped their quays.

The wind is rising, pushing hard from the south-west, rippling the creek as it snakes out through the sandbanks at the river mouth. Thirty or so dark little brent geese are standing there. A curlew spots my moving shape and flies up with a bubbling complaint from the sand into which it has been digging its long, downcurved bill. The obsolete trades of this coast have left their spoor in the scribbly maze of the marsh creeks – ancient fishing boats reduced to mud-embalmed skeletons, a stove-in wildfowling punt minus its ten-foot-long gun, abandoned oyster beds marked out with old wooden stakes still connected by sagging wickerwork panels.

A snaking duckboard trail leads out through the sand dunes between crisp mats of mosses and crusty grey lichens. Hawthorn bushes lean out of the sandhills, their lichened twigs whistling with wind. Yellow horned-poppies bob their big papery petals. Marram grass seethes like a crazy man's hair. The sand of the dune slacks is full of minuscule snail shells, each whorl scoured to a pale ghost by the rushing sand grains. The flanks of the dunes have been sculpted by the wind into countless overlapping layers, as finely as by any palette knife. Swinburne might be a bit too free with the alliteration, and too prone to indulge in pleonasm (a sort of poetic verbal diarrhoea), but two of his verses in particular nail it all very precisely:

> Tall the plumage of the rush-flower tosses,
> Sharp and soft in many a curve and line
> Gleam and glow the sea-coloured marsh-mosses
> Salt and splendid from the circling brine.
> Streak on streak of glimmering seashine crosses
> All the land sea-saturate as with wine.
>
> Out and in and out the sharp straits wander,
> In and out and in the wild way strives,
> Starred and paved and lined with flowers that
>   squander
> Gold as golden as the gold of hives,
> Salt and moist and multiform . . .

Down on the strand the wind forces me along with a hand in my back. On the tideline, now at lowest ebb, delicate little sanderlings dart among the wavelets in a rippling group. A girl runs ahead with her dog, making it leap for driftwood, her long black hair streaming out before her. There's a kind of ecstasy in walking here at this moment, out among the sand dervishes in a shoving gale, lord of a strand half a mile wide, crunching across an epic graveyard of razorshells into a twilight full of the sting of salt and the tang of resin from the pine trees on Holkham Meals.

Opposite Holkham Park I turn inland and walk up Lady Anne's Drive towards the lights of the Victoria Hotel. Thousands of pink-footed geese have flown in from the beet fields to roost on the pools and grasslands of the freshwater marshes below the coast road, and more come past me with their flaps down as I stop and stare. A few years ago some of the geese began to use these marshes as a night roost in preference to the mudflats, where the incoming tide might necessitate a shift of roost halfway through the night. The marshes are well sheltered, too, by the thick belt of trees on Holkham Meals, and wildfowlers aren't allowed to shoot here.

I'm tempted to turn aside and spend an hour watching the spectacle from the bird hide on the marshes. But I know the hide will be packed with birdwatchers, and I don't want anyone else's company just now – not after the solitary magic of the windblown beach. So I stay by the edge of the roadway with my elbows on the fence, mooching and dreaming and rubbing my sore ankles, while my stupidly sensitive feet grumble at me to hurry the hell up and get us out of these bloody boots and into some nice comfy trainers.

My father's feet, like mine, were too sensitive for his own comfort. All his walking life they plagued him. Some tropical fungal nasty that he picked up during the war turned his toenails yellow and rotten. He dusted his socks with medicated talcum from a special blue and yellow tin. He winced when he cut his nails. After walks his feet were patched with plasters. He didn't complain, of course – just bore it all, stoically.

As far as walking boots were concerned, Dad was no stick-in-the-mud. He had no qualms about jettisoning his hobnails in the 1950s and embracing the lighter, more comfortable Vibram soles. In the matter of domestic footgear he remained a little more conservative – black lace-ups for the office, brown for home, both types polished to a high gloss. But he was a natural fool for Dr Martens' high-sided, lightweight boots when they hit the outdoor shops of Gloucestershire in the early 1960s. Round the house, down the lane, up in the attic, out in the garden, for the next forty years Dad sported his Docs with their yellow-stitched AirWair soles, stolidly stamped 'Oil, Fat, Acid, Petrol, Alkali Resistant'.

The Docs wore out (he literally walked their soles flat), and had to be replaced. Black brogues beyond repair gave way to more black brogues, superannuated brown clumpers yielded to identical successors. But one elegant pair of shoes breasted the years like true survivors, making periodic appearances on Dad's feet, then retiring to 'rest between engagements', neatly aligned side by side like pensionable actors behind the miniature safety-curtain of his shoe shelf – the Malta shoes.

Dad had the shoes handmade in Malta when he was serving in the Mediterranean in the early stages of the war, and they lasted him the rest of his life. Unlike Murphy's Spade with its 'five new blades and seven new handles', the Malta shoes remained part-authentic. The soles were replaced five or six times at least. But the uppers were those of the original shoes of 1940, expertly patched by shoe-menders at various times, worn almost through at pressure points created by the knobbles and knuckles of Dad's feet, and polished to tissue thinness. How he must have loved those Malta shoes. He maintained them for more than sixty years with the pride and care one reserves for something special, an irreplaceable link with old times, like a friendship.

I look at my own shoes now, a pair of cheap old knock-abouts with frayed laces and a hole in one sole. I don't care about them, and I don't care for them. They haven't seen the polish brush for weeks. I don't exactly blush for shame, but I feel the stretch of the years and the changing times between father and son.

Shoe polishing was a gentleman's obligation, another in the long list of ritual chores that a man didn't shirk. The polish came in little round tins labelled 'Kiwi', with a picture of a furtive-looking bird on the lid. There was a special lever at the side, and as a child it was a long while before I mastered the knack of twisting and pushing it at the same time to make the tin pop open, emitting a tiny sigh and a pungent breath of shoe polish. Sometimes the lid would fly off and roll across the floor with a clatter. Hairs from the shoe brushes (one to apply the polish, the other to work it to a shine) stuck to the greasy mixture in the tin. Over time the polish would decompose into a substance like cake crumbs which defiled anything it touched with tiny specks that grew huge if you smeared them with your finger.

Dad's Malta shoes were exotic. They were objects of

mystery and wonder to me as a boy. The application of Kiwi tan polish and elbow grease, year in and year out, gave them a subdued but intense shine, and a kind of mottled patina of a colour that was unique, but which I now find impossible to describe. They were going-to-church shoes, even in Dad's old age. And he brought them with him when I took him back to Malta shortly after his eightieth birthday, for what turned out to be our last walking holiday together.

We explored the streets of Floriana, where Dad had lived at various times during the 1920s and 1930s when his father was stationed in Malta with the Mediterranean Fleet. The tremendous view over the Grand Harbour brought back lots of memories – the notice prohibiting spitting on the bus ('Ma bżiq fuq ix-xarabank') at which Dad and his sister Rachel used to snigger, picnics and walks on the Dingli cliffs, swimming jaunts at Tigné, and visits to the big ships for Christmas parties at which balloons were blown by compressed air out of the muzzles of the great 15-inch guns.

One afternoon we wandered along Strait Street, a narrow thoroughfare tremendously long and only four yards wide, known to British sailors of my father's day as The Gut, where red-light bars and dance halls catered for the Royal Navy's lower deck all through the war. Now the Smiling Prince, the Union Jack and the Blue Peter crouched dusty and shuttered in the shadows. Only Miss Lulu at No. 101 seemed to be still in business. 'Officers wouldn't have been seen here,' murmured Dad as he contemplated the faded sign of the Splendid Lounge Bar. 'The Gut was where the sailors would come to drink far too much beer, and get a bit of slap and tickle.'

It was as he turned away from the Splendid Lounge Bar that Dad suddenly staggered sideways and cannoned into the wall like an able seaman on the spree. A couple of days before, walking on the neighbouring island of Gozo, I'd noticed that he had been finding it difficult to lift his feet in walking boots. He'd stumbled on rocks he'd normally have stepped over without noticing, and eventually he'd fallen, barking his shins and letting out a clipped, officer-like '*Fuck!*' Since then I'd seen him moving with an unsteady, lurching gait along the pavements of Valletta. Without warning the footpaths had become an obstacle course studded with the raised edges of paving stones, foot-high front steps, café chairs and – deadliest of all – the almost invisible cast-iron rings of lamp standards that had been sawn off an inch above pavement level. I found myself holding my breath wherever we walked, waiting for him to fall, reaching out a hand by reflex action every minute or so, then withdrawing it again in a nod to his dignity. Outside the Splendid Lounge he cursed himself under his breath: 'Silly old idiot!', and turned away from my hand under his elbow. Even if he needed help, he didn't ruddy well want it. It was my first proper intimation of what it meant to take the reins, of the role reversal that old age visits on offspring and their parents, the child learning to act as father to the man in yet another twist of that versatile aphorism.

There were no more companionable walking expeditions, home or abroad, after the Malta trip. Dad's confidence had been shaken. He didn't want to 'be a bore', he didn't want to 'hold everyone back', he hated the idea of toddler walks through the Bordeaux vineyards or down the Thames. If he couldn't tackle a tough mountain any more, that was fair enough at eighty-something. But now that hill tracks and stony paths were obstacle courses, now that pebbles had become problematic and irregular pavements lay in wait for his uncertain steps – well, he wasn't going to settle gracefully for being the object of anyone's solicitude. He was the manager, the man in the driving seat, wasn't he? None of this bloody pipe-and-slippers walking, thank you!

Eight o'clock in the morning, and a kennel of hound puppies at full cry in the sky over the Victoria Hotel. That's what it sounds like, anyway, as I limp to my bedroom window and pull back the curtains. Four thousand? Five? Impossible to count the pink-footed geese as they stream overhead towards the beet fields and stubbles inland. The low grey sky is packed with their bulky bodies and stiffly sawing wings, and full of the squeaks and yelps by which the members of family groups within the main pack communicate with one another. I linger at the window, mesmerized by the noise and action, until the last honk and whistle of wings have faded to the south.

Two more days eastward along the Norfolk Coast Path are ahead of me, if the feet will allow it. I go through the tedious lean-and-stretch exercises that plantar fasciitis sufferers have to force themselves to do, and my heels moan and bitch at me until the stretching shuts them up.

Down on Holkham Meals the pines sway and roar. The dunes they shelter lie quiet, and so does the strip of waterlogged salt marsh beyond. The creeks are full of tiny, pale grey waders with black beaks – sanderlings, scurrying along the muddy windings before lifting off in a twinkling mass. A clap and thump of hammers heralds Wells-next-the-Sea, where beach-hut owners are battening down the hatches ahead of the winter's storms. The huts stand in line, green and blue, puce and purple, with little balconies and staircases and fretwork gables. Essentially they are clapboard sheds, but they are also palaces of dreams, and they change hands for up to £70,000.

A straight mile-long channel through reclaimed land and salt marsh connects Wells beach with Wells town. The task of keeping it clear of silt falls to the Port of Wells's stout little dredger *Kari Hege* and the JCB she carries. Alongside the flood bank *Kari Hege* grunts and rattles. The JCB on her deck pokes a long, double-jointed beak into the water like a giant yellow wader digging for worms. It brings up a dripping bucketful of silt, which it deposits surprisingly delicately on a mudflat at the far side of the channel.

Wells is another of North Norfolk's coast towns that have been cut off from the open sea by the spread of the salt marshes. A beautiful old Dutch sailing clipper, *Albatross*, lies moored at the quayside. There's the jealous *raw-caw!* of herring gulls, and a sudden rushing flight of turnstones across the harbour. Two elderly men are riding bicycles side by side – not sweating members of the Lycra brigade, but proper old cyclists in flapping coats and gumboots, sitting upright as though on horseback, pedalling steadily and wordlessly with knees akimbo. I follow them along the waterfront until they branch off into Jolly Sailor Yard.

Beyond East Quay the coast path resumes its eastward march along the margin between farmland and salt marsh. The wind swirls a heavy, miasmic mist across the sun. An oystercatcher stands mirrored in the glistening mud of a creek, watching my approach. It opens its bulky orange bill and emits a sharp *pik! pik!* before flying away in a black and white vee towards the distant sea. Hereabouts the salt-marsh carpet stretches away from the land for the best part of two miles, its sea lavender faded from rich purple to a hazy mauve. The map suggests that the outer edge of the marshes is drawing nearer the further east I walk, but there's no visible evidence of that. Looking north from the coast path, the seaward horizon still seems dead flat and dun brown. There are wavy lines of geese far away, although the heaviness of the morning sky and the monotonous tone of the salt marshes rob them of definition. My feet are aching, and the beer I drank in the taproom of the Vic last night has lodged a complaint somewhere at the back of my skull. But the downbeat appeal of this windswept, salty landscape draws me on east, along the marsh rim and up a muddy lane through the knot of trees that hides the little village of Stiffkey.

'A good man, and has been much maligned,' says the woman I find placing a wreath in the churchyard. 'Perhaps a bit naïve, though.' Stiffkey will never be rid of the lurid tales that swirl around its pre-war rector Harold Davidson. For more than thirty years this diminutive parson, a comic actor before his ordination, pursued a self-imposed mission to save the souls of London prostitutes. He carried out this ministry with an energy he failed to display in his church duties. He poured out money he couldn't afford on buying the girls meals and paying for their rooms. He would spend days and nights with them in Soho, and invite them to stay mob-handed at the rectory in his very conservative Norfolk parish. It seems unguarded behaviour to say the least, and eventually it all blew up in his face. In March 1932 proceedings began against him in a consistory court of the Church of England, and in October that year he was formally defrocked for immorality. After his disgrace, to make ends meet, Davidson turned to public performance in Blackpool, exhibiting himself in a barrel on the Golden Mile. His end on 28 July 1937 was as theatrical and tragic as his entire story seemed to demand. Playing the part of Daniel in the Lions' Den on Skegness seafront, he was mauled by Freddie the lion in full view of an audience, and died in hospital two days later.

All the evidence points to a sad injustice having been inflicted on Harold Davidson. The main case against him was based on an accuser's letter of dubious provenance, which his lawyers failed to challenge. The probability is that the church authorities took the opportunity to get rid of a troublesome, obsessive man with a stubborn streak, an awkward character who was bound to rub such a conservative institution the wrong way.

Davidson wasn't the only man with uncomfortable personality traits to leave his mark on Stiffkey. Henry Williamson was another lifelong member of the awkward squad. In the same year that the errant rector met his end, the author of *Tarka the Otter* brought his family to live in the village and immediately made himself unpopular.

Between the ages of ten and fourteen I loved Henry Williamson's books more than any others. I lived and died with Tarka in the country of the Two Rivers, lived and died with Salar the Salmon in the estuary and head-waters of Taw and Torridge. I trembled for Brock the Badger as the whisky-soaked diggers drew near his lair; I ran breathlessly over the moors beside Stumberleap, the great red stag, as he fled from his implacable nemesis Deadlock the hound. Williamson had lived and breathed the lives of these creatures, not tucked up in bed with a book, but out in the woods and up in the rainy heather, day after day and night after night. You could believe what he wrote, absolutely and completely. He brought otters and stags, peregrines and salmon to proper life. You could hold them in the palm of your hand and marvel at their every move, here by your pillow in the light of a lamp. You didn't know about Williamson's mental tor-ment, his life-changing experiences in the trenches of the First World War, his political naivety, his social isolation, the things he put his family through, his salutation of Adolf Hitler as 'the great man across the Rhine, whose life-symbol is the happy child'. You found all that out later on.

After a fifteen-year sojourn in the West Country which inspired all his greatest nature writing, in 1936 Williamson decided to up sticks and move to the opposite side of England. His purpose – inspired by idealism and his often-stated admiration for the British Union of Fascists under Oswald Mosley – was to renovate and bring into useful production the 250-acre Old Hall Farm at Stiffkey, which had fallen into dereliction during the general farming slump of the interwar years. The BUF's agricultural policies, a very 1930s mishmash of agrarian romanticism and authoritarianism, were for higher wages for the worker, an abolition of all foreign food imports, and local production increase strictly planned and supervised by district councils. Townsfolk would flock back to the healthy life on the land from which they had been seduced by industrialization, the British would grow their own food for themselves, and the British Empire would become proudly self-sufficient in agriculture. Henry Williamson bought into that vision lock, stock and barrel as he tackled the steep, roadless land at Old Hall Farm with its overgrown fields, neglected watercourses and tumbledown buildings.

Old Hall employed three farm workers, but even with their manpower the task of rescuing a large, derelict farm was gargantuan for an inexperienced, self-taught amateur, especially at the outset when the farm was entirely horse-powered. Williamson worked himself and everyone round him into the ground. The writer's family were all pressed into service, his wife Loetitia and five children (mostly under ten years old) working from dawn till long after dusk, Williamson driving himself harder than anyone, becoming impatient and explosive at setbacks or perceived 'slacking' on the part of his children, growing short-tempered and obsessive over details. This stubborn irascibility, allied to his well-known Fascist sympathies, did him no favours with suspicious locals and not-very-imaginative military officials when war was declared in 1939. Williamson refused to put up blackout, met billeting officers with objections and complaints, and painted a large BUF lightning-flash logo on his street wall. The powers-that-be responded by giving him a taste (albeit a brief one) of the police cells at Wells-next-the-Sea. Local rumour branded him a spy; his neighbours cold-shouldered him. After the war Loetitia divorced him. Their eldest son rebelled against his father's autocratic regime, gave him a thumping, and emigrated to Canada. Williamson sold up and retreated back to Devon to lick his wounds. He left Old Hall Farm, ironically, in tiptop shape, those painful and destructive eight years of labour having restored the land to A1 grade.

On the road rising inland from the village I find a view-point looking towards Old Hall Farm under its wooded ridge. On the far side of the River Stiffkey, the turreted walls of Stiffkey Old Hall stare right back. The land around is steep and difficult, pale alluvial soil full of flints, stubborn land whose stubbles and plough furrows glint with standing rainwater. You can hardly imagine a more suitable landscape for an angry, driven man to test himself and his dependants to destruction.

I recross the coast road and follow the rutted lane of Bangay Greenway down to the shore once more. The salt marshes stretch out flat and grey in the watery afternoon light. My shadow points the way I need to go, but I'm tremendously tired. I shift my sore feet from side to side, thinking of the Coasthopper bus. Morston and Blakeney, Cley next the Sea and Salthouse – that's my route east. It's the bus route, too, and those are the villages with the beds and the breakfasts. Cut the walk short? Take the bus? What would Henry Williamson think of that? I stand on the marsh path, weighing my options, as a dozen pink-footed geese come into view. They come steadily nearer, and I curb my hopping and stand stock still. Intent on their own purposes, seeking safety from fox and man, the geese pass right over my head and go yelping and whistling away towards the invisible sea.

November

*The poor November man sees fire and
wind and mist and rain and Winter air . . .*

A COLD NOVEMBER WIND in a hanging grey Wiltshire sky, and I am expecting to find all change along the Harroway. I'm convinced a raw new housing estate at Weyhill will have arisen where the fair sheds used to stand. But the old sheds are still there. I smile with pleasure to see their white walls stretching away along the roadside. Last time I was here, on a scorching May day twenty-five years ago, an unbroken stream of traffic had swept past these long sheds, the heavy westbound lorries shaking the hundred-year-old cob from their walls and spattering them a filthy grey with gutter spray. Now Weyhill lies bypassed and becalmed, and the A303 roars like a sulky lion two fields away.

It feels strange to be sauntering into the courtyard of the craft centre that has housed itself in the fair sheds, to get a cup of tea from the Ewe And I tearoom and read the history of Weyhill Fair from informative plaques displayed around the yard. On that early summer's afternoon when I first pitched up here, no one had seemed to know or care about the old sheds. I didn't care much about them myself. I was sunstruck and footsore, and absolutely done in. I'd walked forty miles in two days from Farnham, dodging back and forth down green lanes and busy roads as I tried to puzzle out the course of the Harroway, the oldest and the most thoroughly forgotten road in Britain. All I wanted was a nice tall drink, with plenty of ice in it.

But the pub was shut and there was no other source of liquid. So I slumped panting against the shed wall, until a talkative old man took my mind off my troubles.

'Heard of Weyhill Fair? No? Hmm, well, nobody has these days. But these sheds here' – he slapped the crumbling cob with the flat of his hand – 'my dad told me they used to be stuffed right up to the rafters with hops, hops in pockets, they were big old sacks, on the fair days come October. All the way down from Farnham on the old road, in big wagons pulled by horses, that's how they brought them. You look at your map, you'll see six lanes meeting at Weyhill. But the old road from Farnham was the one. Dad said it would be so full of sheep coming to the spring fair that you couldn't see the front or the back of them.' He rapped his stick on the road. 'I saw the fairs when I was a boy before the war, but they were nothing to what my dad talked about. He saw cheeses sold by the ton from these sheds, hops by the hundred ton. Fighting and drinking, gypsy horse sales and races, dogfights behind the tents. Lots of rough 'uns about in those days, boy! And I expect you know about the man who sold his wife at Weyhill.'

I must have looked blank, because he snorted with surprise. 'No? Well, have a look upstairs in the pub when it opens. Plenty there to keep you interested!'

The acrylic colours of the mural in the upstairs room of the Weyhill Fair Inn are almost as sharp today as they were when I first laid eyes on them. It's a tremendous piece of work that covers an entire wall. Here are long-horned cattle and downland sheep being driven along the

Harroway to Weyhill. A bareback gypsy gallops a white horse, showing off its paces. Men in shirtsleeves unload bulging hop sacks from high-piled wagons while buyers slit open the hessian to test the contents. And there is the tent with its notice proclaiming 'Good Frumity Sold Here', where candlelight shadows the faces of a crowd watching as a sturdy man in a sailor cap lays down notes of money to purchase a tearful young woman. Thomas Hardy's rash Mayor of Casterbridge, Michael Henchard, got drunk on frumity – broth of boiled wheat and milk, laced with rum – and sold his wife at 'Weyford-Priors' fair. But Hardy got his inspiration from an incident at the Weyhill Fair of 1832, when pensioner Henry Mears paid twenty shillings to farmer Joseph Thomson for his wife of three years, Mary Anne, 'a spruce and lively damsel, apparently not exceeding 22 years of age'. The *Annual Register* for 1832 records the transaction as having taken place in Carlisle, but as the names and circumstances are identical the account must be of the Weyhill wife sale. Mary Thomson, reports the *Register*, stood on an oak chair with a straw halter round her neck while her husband presented her to prospective purchasers in these terms:

> Gentlemen, it is her wish as well as mine to part for ever. She has been to me only a bosom serpent. I took her for my comfort, and the good of my home; but she became my tormentor, a domestic curse, a night invasion, and a daily devil. Gentlemen, I speak truth from my heart, when I say, may God deliver us from troublesome wives

and frolicsome widows. Avoid them as you would a mad dog, a roaring lion, a loaded pistol, cholera morbus, Mount Etna, or any other pestilential phenomena in nature.

Now I have shown you the dark side of my wife, and told you her faults and her failings, I will now introduce the bright and sunny side of her, and explain her qualifications and goodness. She can read novels and milk cows; she can laugh and weep with the same ease that you could take a glass of ale when thirsty: indeed, gentlemen, she reminds me of what the poet says of women in general –

Heaven gave to women the peculiar grace,
To laugh, to weep, to cheat the human race.

She can make butter and scold the maid, she can sing Moore's melodies, and plait her frills and caps; she cannot make rum, gin, or whisky; but she is a good judge of the quality from long experience in tasting them. I therefore offer her with all her perfections and imperfections, for the sum of 50s.

After an hour with no bids, Henry Mears stepped in with his offer of twenty shillings. He must have been pleased with his side of the bargain, because he threw in a Newfoundland dog as well. 'The happy couple immediately left town together, amidst the shouts and huzzahs of the multitude, in which they were joined by Thompson [sic], who, with the greatest good humour imaginable, proceeded to put the halter, which his wife had taken off, round the neck of his newly-acquired Newfoundland dog, and then proceeded to the first public-house, where he spent the remainder of the day.'

Weyhill Fair could be a rough occasion, but it throve for at least seven hundred years, probably much longer. Half a million sheep were sold each year in the heyday of the fair, 750 tons of hops, and doubtless a good number of wives. It was a very male environment, with its own rituals and initiation ceremonies. One such, at Weyhill's October Hop Fair, was 'Horning the Colts'. Anyone might find himself a candidate for initiation – an under-carter wanting to better himself, an apprentice shepherd keen to impress his master, or an unsuspecting stranger. Even a vicar newly appointed to the benefice was not exempt from being lured into the horning ceremony at the Star Inn. After a good dinner all round, the landlord would rise and call for silence before demanding, 'Any colts to be horned?' A barbarian crown was then brought forward. It consisted of a pair of ram's horns mounted in silver, with a silver pint cup full of beer fixed between them. Below the cup a silver gargoyle leered, open-mouthed. The crown was lowered onto the colt's head, while the company sang lustily:

As fleet runs the hare, as cunning runs the fox,
Why should not this young calf live to grow an ox?

For to gain his living through briars and through
     thorns,
And to die like his daddy with a long pair of horns.
Horns, boys, horns!
Horns, boys, horns!
Why shouldn't he ramble amongst briars and thorns,
And to die like his daddy with a long pair of horns?

The colt was then obliged to drain the silver cup without pausing for breath or spilling a drop. Failure incurred a penalty of paying for drinks for the company. Success saw the drinks paid for by general whip-round – beer for the carters, wine for the well-to-do farmers. As there might be several initiates at a single horning, the ceremony could get extremely drunken and wild.

Horns for fertility, horns for cuckoldry; horns for kingship, horns for misrule. And horns by the thousand, as the inn's mural depicts, on the cattle and sheep that choked the roads at fair time. Nearly a dozen lanes and tracks converge on Weyhill. They brought men, beasts and goods from Somerset and the West Country, the Wiltshire and Hampshire downs, Sussex and Kent, and further afield. But the most important was the oldest of all. If it hadn't been for the Harroway, the mighty old trackway that curves across the chalk highlands of southern Britain, Weyhill Fair might never even have come into existence.

Trackways from the distant past are my favourite walking routes. There's something vigorous and direct about the way they forge ahead across the land. Fences, roads, farm-yards and gardens are crossed as though they don't exist. Housing estates and business parks are shrugged aside. A lane swerves too close and is kidnapped, force-marched along for a mile or two, then pushed away. The old ways pass through the landscape as ghosts pass through walls, transcending barriers of time and space. They remain all of a piece across the Ordnance Survey maps, though they are constantly shape-shifting from green lanes to dual carriageways, suburban roads to woodland pathways. That's how I first became aware of the Harroway some thirty years ago; initially on an Ordnance Survey map of the Hampshire downs, in the shape of a track labelled 'Harrow Way' in the Gothic font which denotes an antiquity, and then as a thread I followed with my finger across a dozen consecutive maps. No one could tell me anything about this mysterious ancient route until, after five years of searching, I stumbled on a densely written, drily factual book called *Ancient Trackways of Wessex* by H. W. Timperley and Edith Brill. This husband-and-wife team had corralled a rich stew of old routes with compelling names – Ox Drove and Ram Alley, Juggler's Lane and Smugglers' Lane, King John's Road and Saxon Drove, Deadman's Lane and Old Shepherd's Shore – a poetry of old roads that called out to me. Across the pull-out map at the back of the book the Harroway ran intertwined among these alleys, lanes and droves, a road that might be seven thousand years old, a conglomeration of tin-traders' tracks and drovers' routes, Saxon warpaths and packhorse highways, an upland thoroughfare for commerce and trade that stretched in a great arc for the best part of three hundred miles between the Channel ports and the Devon coast at Seaton.

I put on new boots that summer, and I followed the old road as best I could. The eastern half, from the Kentish coast to Farnham in Surrey, was now the North Downs Way national trail, thoroughly documented and well walked. It didn't interest me. What I was after was to unmask the nebulous end of the Harroway, to unpeel the imprecise old trackway from its western landscape by map and compass, Timperley and Brill, sense and nose. It took me a couple of weeks to walk its hundred and fifty baffling, well-disguised miles. My feet in their brand-new boots received a battering that almost led to abandonment of the adventure. By the time I got to Seaton I was sick to the back teeth of the Harroway. But it continued to possess my imagination. I told myself I'd come back one day to see the old fair sheds at Weyhill one last time before their inevitable demolition. And I'd walk again the section that most haunted my mind, the twenty-five miles of rutted white trackway over the rolling Wiltshire downs from Stonehenge to the wooded ridge above Stourhead where Alfred Tower raises its triangular height among the beech trees.

Outside the Weyhill Fair Inn I turn my face to the heavy evening sky. There's a wintry pinch to the wind, a threat of sleet showers thickening in the west over the downs where Stonehenge stands. It's proper November weather, Dave Goulder weather, all 'wind and mist and rain and winter air'. Exactly what I've been hoping for, in fact. There is something wintry about the Harroway, its deeply scored ruts and dark enclosing hedges, that is definite, but hard to delineate. Mud is lovingly hugged, rain jealously hoarded by this old trackway. I know I'll be chalky muck to the knees by this time tomorrow; wet, too, and chilled to the bone. But I'll have come closer to those old travellers, the gritty journeys they undertook along the Harroway, than in balmy weather and easy going.

A shivering wait in a wet and knife-keen wind next morning outside the Stonehenge ticket office, then a clattering ride in a bus along what was till recently the A344 main road past the Stones. Decades of controversy and complaint have seen the notoriously intrusive road re-routed, its old course now a grand avenue leading to the monument. Indifferent Japanese students point their phone cameras out of the rain-spattered bus windows and snap the woods and road verges for want of anything better.

Britain's greatest archaeological monument comes up over the eastern skyline quite modestly. As we disembark, I'm feeling sulky about having to share my precious visit with a crowd. But when I have the Stones right under my hand, so to speak, all that bad humour falls away. They do not exactly make things better, but I am forced to agree with what poet-priest Layamon wrote eight hundred years ago in his epic poem *Brut, or The Chronicle of Britain*, a wild and whirling account of the 'history' of these islands:

The stones are great,
And magic power they have;
Men that are sick
Fare to that stone;
And they wash that stone,
And with that water bathe away their sickness.

Seen close to, the Stones look as majestic, as big and weighty as ever in their shaggy grey coats of usnea lichen. The tall 'doorways' of the sandstone trilithons and the tight circle of the bluestones exert their peculiar force. How rough and warm they used to feel against the cheek and palm, in the days when you could walk among them looking for sign and symbol. Nowadays the understated barrier of the roped-off pathway enforces respect, even deference, but there was a sort of affection you could feel for the Stones that I don't sense now.

A strip of trackway to the west of the Stones is the start of my Harroway pilgrimage. A young tour-bus guide waits for her charges there. Rooks and jackdaws hop around her feet. She is feeding a starling on a fence post by hand, placing elderberries one by one into its open beak. 'My favourites, the starlings,' she smiles, 'very sweet birds.' Nearby a hippy bus in faded green paint is parked in the flooded ruts of the Harroway. A tiny Christmas tree sparkles in its front window. In the passenger seat a woman in a huge home-made sweater plies her knitting needles with an air of sublime contentment. She looks rooted, as though she has settled down to knit away the days until the druids arrive to set the ceremonies of the winter solstice in motion.

On the verge of the A303 I skip from foot to foot, awaiting my chance to dodge across the double death-stream of cars and lorries. On the far side the Harroway forges south-west over the green acres of Normanton Down. Here the force field of the Stones subsides in wave upon wave of burial mounds – long barrows, bell barrows, disc barrows, saucer barrows. Their humps are scattered in the pastures like creatures burrowed in for hibernation. Modern archaeological methods of scanning the ground and what lies beneath it have produced a revolution in the interpretation of Stonehenge. It's all about the wider view now, with the Stones regarded not only as one individual monument, but more importantly as the hub – or perhaps one hub among several yet to be identified – of a Neolithic ritual landscape developed over thousands of years and unrivalled anywhere in the world.

Birmingham University has initiated a Stonehenge Hidden Landscapes Project, using ground-penetrating remote sensors to look beneath the skin of the countryside around Stonehenge. Already they have located some exceptional sites, including shrines, temples, ceremonial avenues, barrows never before noticed, burial pits and the ring-shaped, embanked enclosures known as henges. It's a vast field of ceremonies. Thousands of people at a time are known to have travelled to the Stonehenge landscape, but for what purpose is open to debate. They must have been overawed by what they found here. Only a couple of months ago the Hidden Landscapes team announced a new discovery underground at a well-known site, the great henge of Durrington Walls a couple of miles north-east of Stonehenge – thirty huge stones, invisible to all but remote sensor technology, that were part of an arc of nearly one hundred which had been set up some 4,500 years ago round the southern perimeter of the henge before being deliberately toppled over and embedded in its structure. Why? What for? Probably we'll never know. The reasoning behind the actions of our ancestors is so hard to fathom. We don't even know the objective behind the assembly of the Stonehenge monument itself between 3000 BC and 2500 BC – the great circular henge set with its ring of timber posts, the importing of eighty massive vol-canic bluestones from the Preseli Hills a hundred and seventy miles away (how on earth did they manage that?), the erection of those sandstone 'doorways'. Was it a giant clock, a calendar, a temple, an arena? We don't know, and we won't know. But we are of that remarkable species that will not stop trying to find out.

I follow the wide pale furrow of the old road past the Bush Barrow, excavated in 1808 by local antiquarian William Cunnington on behalf of that gentlemanly barrow-digger, Sir Richard Colt Hoare of Stourhead. Cunnington unearthed 'the skeleton of a tall and stout man' whose thighbone measured twenty inches in length; that would have made the dead man a six-footer, quite a giant for his day in 2000 BC. Also in the tomb were a diamond-shaped breastplate of gold beautifully worked with lozenges and zigzags, a bronze dagger and spear-head 'full 13 inches long, and the largest we have ever found', along with bone rings, bronze rivets, and a mace head carved from fossil-rich rock 'composed of a mass of seaworms or little serpents'. What a treasure! It is no use lamenting the detailed story which could have been extracted from such a site if it had been excavated with the aid of modern technology by modern archaeologists. These fabulous items – now displayed in the Wiltshire Museum, Devizes – dazzled and amazed the enquiring minds that did unearth them two centuries ago.

The Harroway passes between a pair of disc barrows, their low circular ramparts perhaps fifty yards in diameter. They were sited as close as possible to the old road, which was probably already a couple of thousand years old when they were built. The large pale pigs who occupy the hillside beyond are busy reducing their field to a uniform sea of mud. They stare at me from under their floppy ears as I pass by, and I think rather guiltily of this morning's crispy bacon sandwich. There is definitely something sardonic about those curvilinear porcine smiles. If pigs were mind readers and that electric fence became disconnected, I'd be a goner.

A plaintive creaking comes from overhead, where fifty lapwings are proceeding jerkily across the sky in a heart-shaped crowd that rises higher and higher until I lose sight of it against the grey clouds. At Druid's Lodge across the Salisbury road, horses put their soft whiskery noses over the paddock fences to be stroked. I love horses, provided they are the other side of a good stout fence. These look glossy and fit. Maybe they would give a fair gallop to one or two of the racehorses that were trained here to remarkable effect by the shadowy Druid's Lodge Confederacy before the First World War. The Confederacy, a syndicate of owners with Irish connections, trailed clouds of mystery in the press and popular imagination owing to their huge levels of success and the air of secrecy that hung around the lonely stables on the old trackway in a hidden fold of the downs. In fact their achievements in winning handicap races with unfancied horses that they backed to the hilt were down to no more than the owners' shrewd eye for a likely nag, an astute trainer in Jack Fallon, and a competent jockey in Bernard Dillon. But to the racing world of Edwardian England there seemed something uncanny and mysterious about the goings-on at Druid's Lodge – a conception no doubt reinforced by the spooky name of the establishment.

Three red kites wheel over the sprawling sheds at Druid's Farm, keeping a gaggle of guinea fowl under observation. The birds waddle hysterically away from me along the Harroway in their grey pinstripe plumage, their scarlet wattles trembling as they shrilly debate the relative dangers of circling kites and advancing human; then all of a sudden they take flight en masse and disappear over the hedge in a flurry of squawks.

The descent into Berwick St James leads past a band of deep and broad strip lynchets driven along the downland slope by medieval ox-ploughs. Down in the valley bottom the old road winds among houses whose walls are chequered in pale chalky stone and dark flint. I cross the River Till, as clear as dimpled glass, long streamers of watercress rippling over its flinty bed. Ancient trackways kept as much as possible to the high ground, out of the mire and vegetation of the valleys. But rivers had to be negotiated every now and then, and the Harroway always makes for a shallow crossing place such as this, with a good firm bottom. A pair of sarsen stones guards the old road's exit from the village under the guise of Langford Waie, an old name if ever there was one.

It's hard to pick a decent foothold on the old track as it climbs between broad hedges. The ruts of Langford Waie are splashy with cream-coloured rainwater from the morning's early showers, and the raised ridges of hard chalk between the ruts shine with a treacherous greenish polish. The bad surfaces of ancient trackways were a notorious problem for previous generations of wayfarers. An Act of Parliament passed in the reign of King Henry VIII for the building of new roads talks of 'many of the wayes in the wealds as so depe and noyous by wearyng and course of water and other occasions that people cannot have their carriages or passages by horses uppon or by the same but to their great paynes, perill and jeopardie'.

A winter journey along such a road was no joke. In former times travellers would move out sideways to circumvent dangerous sections of a road, and there are stretches of the Harroway where the scars of such detours show up hundreds of yards away on either side. But the Enclosures of the eighteenth and nineteenth centuries saw the old ways confined within the bounds of parallel hedges, a situation from which it's impossible to escape the bad going unless by trespass into the adjacent fields. Struggling along, I skid on the polished chalk and come down hard, bashing my elbow and bruising my knees. Ripe words of discontent follow, but there's only a jay within earshot, and he's more than ready to outswear me.

I pick myself up and limp on uphill and out of the trees. Langford Waie ribbons ahead along the shoulder of the downs. A hare springs up right under my feet and sprints off across the winter wheat in fast equine leaps. It bolts over a shallow crest and swerves hard left as soon as it's out of sight, a ruse betrayed by the black tips of its ears which I can see bobbing along the skyline for a few seconds before they, too, vanish.

I cross an army road whose notice, warning of oncoming tanks, is a reminder that the vast military training ground of Salisbury Plain lies not far north of here. Beyond the tank road the Harroway, now under the name of Berwick Lane, begins a long downward slope towards the Wylye valley. The birds have plucked the brambles clean of all their blackberries, leaving a jungle of purple leaves from which viciously barbed tentacles reach halfway across the lane to hook a coat sleeve or a cheek. A vestigial apple tree leans out of the bank, cut and slashed by a farmer's flail to within an inch of its life, its autumn crop of fruit in a rotting pile at its foot. The heap has been colonized by slithery grey slugs, but those apples that remain uneaten have been hacked into deep narrow slots. I find the culprits further down the slope, a flock of fieldfares that has taken over an abandoned orchard. A few birds are bouncing through the bare branches, calling *chic-chic* like squeaky clocks, but most are strutting the grass among the windfalls. The apples have been softened by rot and offer a succulent feast of sweetness to the sharp, stabbing beaks of the fieldfares. These Viking visitors, over from Scandinavia for some winter pillage among the orchards, are such handsome creatures with their pale spotted breasts, russet wing covers and smoky grey heads. Watching them carousing, I'm reminded of the preternaturally observant Richard Jefferies – Wiltshire born and bred – and his description in *Round About a Great Estate* of the order in which the autumn fruits are eaten by the birds, as true today as it was in 1880:

> The first berries to go as the autumn approaches are those of the mountain-ash. Both blackbirds and thrushes began to devour the pale-red bunches hanging on the mountain-ashes as early as the 4th of September last year. Starlings are fond of elder-berries: a flock alighting on a bush black with ripe berries will clear the bunches in a very short time. Haws, or peggles, which often quite cover the hawthorn bushes, are not so general a food as the fruit of the briar. Hips are preferred; at least, the fruit of the briar is the first of the two to disappear. The hip is pecked open (by thrushes, redwings, and blackbirds) at the tip, the seeds extracted, and the part where it is attached to the stalk left, just as if the contents had been sucked out. Greenfinches, too, will eat hips.
>
> Haws are often left even after severe frosts; sometimes they seem to shrivel or blacken, and may not perhaps be palatable then. Missel-thrushes

and wood-pigeons eat them. Last winter in the stress of the sharp and continued frosts the greenfinches were driven in December to swallow the shrivelled blackberries still on the brambles. The fruity part of the berries was of course gone, and nothing remained but the seeds or pips, dry and hard as wood; they were reduced to feeding on this wretched food.

Behind me the easterly sky cracks with a gleam of sunshine, but the cold November wind is blowing more wet weather up from the west. It's a race into the Wylye valley between the rain and me. Notwithstanding heavy going that even Bernard Dillon of Druid's Lodge might have been slipping around in, I run into Steeple Langford the winner by a short head. In All Saints church I wait for the shower to cease rattling its sleet against the windows. A carved stone head on the wall, its mouth worn clean away, looks at me with wide empty eyes from under a short choirboy fringe. I take a step backwards and stare, tilting my chin as I try to read the expression that the mason must have intended, and I'm suddenly and poignantly visited by a shadow man alongside me who mirrors my every move.

Following his struggle with the rubbly paths of Alto Aragon and the diminution of his capacity to keep up with fellow walkers half his age, Dad had pretty much decided that his walking days were over. After the Malta trip and the tumble he took there at the age of eighty, he made up his mind and stuck to his decision. It took three or four years of prevarication on his part, of brochures delivered to his house that remained unread, of booked trips cancelled at short notice, before the penny dropped with me. He didn't want to be a burden or a bore. He didn't want ever again to put himself in the situation of being patched up or clucked over, of having younger walkers sighing as they waited for him at the top of the hill. No matter how tempting the lures I cast towards him, the old sea-devil was not going to rise to them again.

I'd blithely assumed that the older men got, the sweeter and fluffier they became, the more inclined to put their feet up and let someone else take over the tiller. Not a chance of that with a man like Dad. He spurred himself harder and harder to justify his existence. Duty drove him to feats of laborious typing at his desk, letters of advice to his children and admonition to slipshod professionals, letters of condolence to bereaved widows of his Dartmouth term-mates, letters of persuasion to councillors and clergymen. Duty had him struggling to his feet while the rest of us were sitting round the table after lunch, tugging on his gumboots for a session stumping up and down the lawn behind the motor mower. If I picture my father in his final years, it's that last image that comes to mind: harried by duty, running on determination, shaking the black dog away from his heels through the therapy of action.

Dad's talk was of arrangements for his children and his wife 'after I've gone aloft', and he gave every indication of looking forward to a bit of P&Q in the celestial crow's nest. He was becoming increasingly deaf, partly as an effect of wartime naval gunfire. It isolated him in social situations, and irritated him enormously. His desk drawers filled up with boxes of different sorts of hearing aids that he wasn't wearing – they were too conspicuous, or uncomfortable, or squeaky, or just damn bloody annoying. I became accustomed to moving my chair directly opposite him so that I could speak slowly and squarely into his face, conscious as I did so of the hair-trigger gap between helpful clarity and patronizing baby-speak. Dad had always been a light eater, very resistant to suggestions of second helpings – 'I'm the General Secretary of NUFF,' he'd tell Mum with a grimly resistant smile, 'National Union against Forcible Feeding.' But as his appetite dwindled in the last year of his life I found myself slipping titbits onto his plate when he came to Sunday lunch – single grapes one by one, a cube of cheese, a square of his favourite chocolate. It was what I had done for my children when they were little; what he had done for me nearly sixty years before.

I don't know whether the irony registered with him. Even now, with the end of his life in sight, I couldn't ask him such a personal question. But I became enormously fond, protectively fond of this father whom I had taken so long to get to know. I found out how much I really loved the old man. I became aware that the reins of our relationship had passed into my hands. I could embrace him now without awkwardness. On greeting and parting I would touch my lips to the side of his head, as much of himself as he'd suffer me to kiss, and smell his musty old hair, and give him as much of a hug as I dared without hurting his frail old ribs.

In the spring of his eighty-seventh year I asked Dad if there was one place he'd like to go back to if a magic carpet were made available. 'Oh, France,' he said without hesitating. 'I'd love to see those gorgeous Romanesque churches in the Auvergne again. I don't suppose I ever shall, though. No, I've said my final goodbyes to them, I should think.'

A few months later we were down there with a list he'd drawn up (researched, as always, from the file of yellowing clippings he'd hoarded since the 1960s), motoring round, ticking off the beautiful churches he loved. He was tottering a lot now, but could still cross a cobbled square with the aid of his trusty ash-plant stick. Wandering the side lanes in second gear, stopping to sniff wild thyme on a hillside or have a glass of wine and a bite of cheese in some sleepy village at three in the afternoon, letting the conversation ramble from the wartime Mediterranean to the bloody awfulness of hearing aids by way of baroque music and scurrilous sailors' songs, Dad and I wound down into the kind of peaceful harmony that back-country France so beautifully induces.

On one of our last days we found ourselves in a church that Dad had never heard of, the abbey of Mozac near Clermont-Ferrand. The carved capitals were the best we had seen. Three of them had been brought down to the floor of the nave so that their exquisite sculptures could be appreciated close to. One of the scenes showed three almond-eyed women at the tomb of Jesus. Clearly moved,

Dad spent a long time gazing at these figures. 'Of all the splendid things we've seen on this trip,' he murmured, 'I think it's these marvellous timeless expressions I'm most drawn to.' He pushed back his head, and I watched him standing motionless, staring and musing, as the carved women gazed back at him across the centuries.

The windows of All Saints church cease their rattling. The sleet shower has marched on east, and it's time for me to leave these shadows. I close the church door softly behind me, as though I might disturb someone, and set out once more along the Harroway.

By the time I get down to Hindon I have seen more than enough of Dave Goulder's 'wind and mist and rain and winter air'. The fire is all that's missing for this poor November man, and as soon as I push open the door of the Lamb Inn I find it, burning like the blazes in a giant open fireplace. A dozen red-faced men are gathered in the room, half of them round a bare scrubbed table by the fire, the others in a tight semicircle at the bar. They are drinking beer and chaffing each other in Wiltshire voices as thick as cream. There's the kind of smell you get after a big pheasant shoot – sweet thick woodsmoke, a sour whiff of cartridge powder, a tang of blood, beer, damp dog and well-sweated tweeds. Half these men are young game-keepers from a rich man's estate nearby, dressed as though by Merchant Ivory for an Edwardian film in tweed plus-fours and satin-backed waistcoats, heather-mixture stockings, tweed caps and check shirts. The others, mostly older men, are in long trousers and thick outdoor jackets. 'Beaters,' grunts the oldest, 'the lowest of the low.'

'Why are the keepers . . . well, why are they dressed like that?'

'Oh, His Nibs likes it, and the clients expect it. And since they're paying the bills . . .'

W. H. Hudson stayed at the Lamb Inn in the spring and summer of 1909 while he was writing his classic rural book *A Shepherd's Life*. Among the very oldest locals, he says, were a few whose memories stretched back as far as the notorious 'Swing Riots' in the district, eighty years before.

> Some . . . were able to recall that miserable and memorable year of 1830 and had witnessed the doings of the 'mobs'. One was a woman . . . now aged ninety-four, who was in her teens when the poor labourers, 'a thousand strong', some say, armed with cudgels, hammers, and axes, visited her village and broke up the thrashing machines they found there.
>
> Another person who remembered that time was an old but remarkably well-preserved man of eighty-nine at Hindon . . . He was but a small boy, attending the Hindon school, when the rioters appeared on the scene, and he watched their entry from the schoolhouse window. It was market-day, and the market was stopped by the invaders, and the agricultural machines brought for sale and

exhibition were broken up. The picture that remains in his mind is of a great excited crowd in which men and cattle and sheep were mixed together in the wide street, which was the market-place, and of shouting and noise of smashing machinery, and finally of the mob pouring forth over the down on its way to the next village, he and other little boys following their march.

Disturbances like these surged in a wave across southern England and East Anglia in the latter part of 1830. They came to be known as the Swing Riots, with many landowners receiving threatening notes signed by a fictitious Captain Swing. They were a terrifying experience for those in their path, but they did not come out of the blue.

By 1830 the lives of the agricultural labourers of England had become pretty much unbearable. The common land on which they had traditionally grown vegetables, grazed animals and foraged had been enclosed and claimed by rich landowners. The landowners' farming tenants, themselves not necessarily well-to-do, had reduced their workers' wages, expecting the Poor Law guardians of the parish to pick up the slack in the form of relief, and the guardians had reduced the level of relief on the grounds that the local tax payers couldn't be expected to dig ever deeper in their pockets. Farmers and labourers no longer worked side by side, and as the social and financial gulf between them widened, so short-term contracts between man and master became the norm – often with

notice to quit being as little as a week. Agricultural workers and their families were poorer, more dependent and
more insecure than they had ever been, just at the time
that labour-replacing devices were being introduced onto
farms. Mechanical thrashing machines which could do
the work of fifteen men were the objects upon which the
rioters of 1830 initially took out their fury, but in truth
the agricultural rebellion of that year had been building,
volcano-style, for decades before it finally erupted.

The aftermath was harsh. Of some two thousand rioters brought before the courts, more than six hundred
were imprisoned, more than five hundred transported.
Sentence of death was passed on 252 of the men, and nine-
teen of these sentences were carried out. The Swing Riots
gave the Establishment a smack across the face, and woke
it up to some of the workers' genuine grievances. The
undue power of rich magnates in their local areas was
soon curbed by two important Acts of Parliament. Two
years after the riots, the Reform Act of 1832 abolished the
notorious 'rotten boroughs', tiny constituencies which
could return a disproportionately influential MP. And in
1834 the Poor Law Amendment Act took responsibility
for the relief of poverty out of the hands of local bigwigs
and placed it in those of specially created workhouses.
But as for the workers themselves, there was little or no
improvement in their rock-bottom situation.

I sit warm and cosy in the Lamb at Hindon, drinking
my beer with the beaters and thinking of what I was told
earlier in the week by a tenant dairy farmer in north
Wiltshire. His herd of a hundred and fifty cattle were in

their winter sheds, it was a couple of hours before milking time, and he had a moment to lean on a gate and talk. It cost him about 22 pence to produce each litre of milk, he reckoned. And the price he was being paid by the co-operative that bought his milk had just been cut from 21 pence to 20 pence a litre, the price that the budget supermarkets were charging their customers. There was no profit for this farmer at all; in fact, he was operating at a loss.

'It's a world market now, of course we know that,' he said, frowning at the thought of it. 'But I shall have to go out of business. My children look at me, pinching the pennies, always working, with the paperwork eating up my evenings, and they say they're damned if they'll follow in my footsteps.'

Shortly before our conversation, this thoughtful and naturally conservative man had been part of a group of farmers that had staged a protest in a cut-price supermarket for the TV cameras, emptying the shelves of milk and leaving a traffic-jam of trolleys laden with the stuff to clog up the checkouts. 'I'd never have thought I was capable of doing such a thing, but what else is there for us but protesting and shouting about our troubles – all the dairy farmers' troubles?' He flicked his hand impatiently, as though batting away a fly. 'It gets me right here in the guts, you know. We are all getting very angry, and that's not right. It's just not right.'

In the morning I wince my way into motion. The A303 roars past at fifty cars a minute, and I'm glad to leave it and follow the Harroway across the shoulder of Charnage Down on a divergent course. The road noise soon diminishes to a whine, and then a negligible murmur. The sky is all bruised cloud, with slabs of milky rain marching on the southern skyline. In the south-west there's a sulky pearlescent gleam low along the hills. The wind is wintry, the land quiet and heavy with rain.

The Harroway runs west, deeply puddled, a diffuse trackway of green grass ridges and pale chalky trenches dug by the heavy-duty tyres of off-road drivers and tractors, their upper edges squeezed proud of the surrounding mud like scar tissue. I pace out the width of the track – fourteen strides from one side to the other of this braided old road dinted by horseshoes, dog pads and boot soles. Everything lies under dull November light, with gauzy moisture seething across a landscape of abrupt chalk ridges and hollows. To the south stretches a great plain where patches of sunlight creep in brilliant green segments among the woods and farms.

When I walked the Harroway twenty-five years ago, this section across Charnage and Mere Downs was measured off in milestones, each stubby stone date-stamped '1750'. Today I can't locate them. Have they been dug up for saleable souvenirs or swallowed by brambles? Perhaps they have been knocked flat and pulverized by off-road drivers skidding by for fun. At Mere Down Farm a man is stringing barbed wire along the fence that borders the Harroway. 'Milestones? Oh, the farmer moved 'em over there.' He gestures vaguely across the field. 'Trying to claim a bit of extra ground, I 'spect.' Half a mile further on

I come across a perfect specimen. It stands by the track, round-shouldered, pocked with lichen and bearded with moss, proclaiming the mileage: 'XXI from Sarum, XCIX from London'. The 1750 date incision is blurred by time and rubbing sheep, but I can make it out. The Harroway provided the main coaching route between London and the West Country before the valley road was turnpiked, and here is the proof in solid stone.

I cross the road to Mere, and come under interrogation by a pair of handsome Rhodesian Ridgeback dogs. I stand extremely still while they investigate the crannies of my body with their large black noses. Their general burliness, their furrowed brows and serious 'hombre' expressions are given the lie by their feet. You can't wear white ankle socks and be a killer. Their owner apologizes, but I don't mind the close inspection. It's a dog thing, and fair enough.

The sky ahead is a swirl of silver and grey. In the track-side fields black cattle munch turnip tops, each mouthful torn free with a small plosive exhalation. Fat black-faced lambs trot away, then turn to stare as though mortally affronted. A flock of several hundred starlings crosses the track in a flicker of black and mauve, their iridescence muted by the low light. They do not come to rest, but stay shimmering just on and above the ground, getting up every minute in a dense, shape-shifting spiral, moving restlessly on by communal agreement. I wonder why, as I have wondered so many times about so many aspects of animal behaviour in the course of this walking year.

Now the flat-topped reservoirs on the summit of

White Sheet Hill are in view ahead. The Harroway wriggles just north of them, its essentially straight course distorted by foreshortening into a series of concertina curves. The next milestone comes up, XXII from Sarum, C from London – one hundred miles from the fabled city that so few country men and women of 1750 ever saw.

A slim figure far ahead on the old trackway turns into Jane, coming east to meet me. At the top of the rise we leave the Harroway and walk the slopes of White Sheet Hill together among the green trenches and ramparts of a Neolithic camp and causeway, an Iron Age hill fort and a twentieth-century water plant. The down is extravagantly steep and curved. Round barrows pimple the summit. Sir Richard Colt Hoare and William Cunnington did their best up here in 1807:

> Immediately on ascending the hill called Whitesheet, we find ourselves surrounded by British antiquities. The road intersects an ancient earthen work, of a circular form, and which, from the slightness of its vallum, appears to have been of high antiquity. Adjoining it is a large barrow, which we opened in October 1807, and found it had contained a skeleton, and had been investigated before.
>
> On a point of land near this barrow are three others, all of which, by the defaced appearance of their summits, seemed to have attracted the notice of former antiquaries. No 1, the nearest to the edge of the hill, had certainly been opened, and appears

to have contained a double interment. The primary one was an interment of burned bones deposited within a shallow cist, in an urn rudely formed, and badly baked. Above it was a skeleton with its head laid towards the south, and which from its position and perfect preservation appears not to have been disturbed. Its mouth was wide open, and it 'grinn'd horribly a ghastly smile', a singularity we have never before met with.

Standing beside the barrow we catch a glimpse of Alfred Tower raising its finger to the low grey sky above the woods of Stourhead, and Sir Richard's mansion glowering in rainsoaked grey down in the vale among more massed trees. The first proper downpour of the day marches in and we pull up our hoods and go sliding down a skiddy chalk path, emerging on a lane that leads to the old drovers' inn at Kilmington crossroads. The Red Lion is the 'Pride of Kilmington', according to its sign, and is 'passionate about beer and food', according to its website. I'd called in on that first Harroway expedition, and to judge by the state of their cheese sandwiches the passion had most definitely been all about the beer back then. 'That old road past the door,' the man at the bar had said, 'my granddad used to drive sheep to Andover on that. A Roman road, it is, bet you didn't know that!' – and he'd nodded complacently like one who had deliberately let slip a secret.

We follow the old road across a heath among amber beeches with wet black trunks, their lower branches still sparsely hung with leaves in gold, green and crimson. In a week the gales that are being forecast will have blown them all into the mud to blacken. Among the trees on the crest of Kingsettle Hill, Alfred Tower rises like a dream tower in a witch's wood. Sir Richard Colt Hoare's grandfather, 'Henry the Magnificent' of Stourhead House, had it built in 1772 to glorify two West Country heroes – King Alfred the Great, and himself. The three-sided red-brick tower shoots skyward, window above window, topped with a candlesnuffer cone; a Rapunzel tower. From the top you can see nearly a hundred miles, from Glastonbury Tor to the Dorset coast.

I came here on blistered feet when I first walked the Harroway. The tower's curator, a darkly humorous man, had led me across the road and into the trees. He poked with his shoe at an ancient round stone, half buried in the undergrowth. Carved lines segmented its rim into three equal parts. 'Could be a plague stone,' he'd said. 'They swapped goods and money across it when the Black Death was on. But some say it's the old boundary stone between Wiltshire, Somerset and Dorset. You stand on it, blow a kiss into your own county, fart towards another, and piss into the one you don't like. What d'you think of that?'

I fossick around for the boundary stone so that I can blow a kiss into Somerset, but it's hidden away in the brambles. I blow the kiss anyway, and walk to the brow of the hill so that I can stand and watch the old road run on down the hill, away on west to Seaton and the sea.

December

December man looks through the snow,
to let eleven brothers know
They're all a little older.

THE LANE IS HALF flooded and muddy as hell. There's one narrow dry strip to walk, and it has been decorated with dozens of dog sausages. God almighty. Someone has done half a job of cleaning up; they have neatly collected a bundle of turds in a plastic bag. Well done! And then they've hung it on the barbed wire fence and walked away. Who do they think is going to dispose of it – the farmer? The council? The pixies?

Sulky reflections for a sore head on Boxing Day morning. Then I turn the crook of the lane and see Cley Hill rising clear ahead, and everything else just falls away.

Ritual. It is an absolute cornerstone of the human condition, this need to note the days in ceremonial ways. Some feel it more strongly than others. I, for one, can't rest satisfied if the turning circle of the year is not properly marked off with home-grown rituals – small ones, but necessary. Sealing the celebrations of Burns Night with a communal round-the-table recitation of William McGonagall's 'The Tay Bridge Disaster'. Ralph Boddy picking out 'He Is Risen' in primroses along the windowsill in Dinder church for Easter morning. Fans and friends of the 'greatest local band in the world', Dr Feelgood, walking the sea walls of Canvey Island together on singer Lee Brilleaux's birthday in May. Joining the gathering of melodeon players on winter Tuesdays round the fire in the Horse & Groom at East Woodlands. Singing carols under the lamp by the old red phone box on our square. Making sure the tattered Chinese lion is in his proper station in the uppermost fork of our Christmas tree. And capping all others, the Boxing Day ritual that starts – rain, shine, snow or blow – with this climb up Cley Hill.

Cley Hill lies in the chalk country at a meeting point of Somerset and Wiltshire. Its demonic origins are beyond question – locally, at least. Every true son and daughter of Somerset knows what a bad lot they are across the border in Wiltshire. The Devil, it seems, was particularly proud of his work in the Wiltshire town of Devizes, until the day when one of his imps let slip that the townsfolk had all gone and got themselves baptized. Old Nick was so affronted that he scooped a great pile of earth into his sack and set off to bury the town and everyone in it.

After a while the Devil's back began to ache with the weight of the sack. He let it drop and sat down for a breather and to ponder some wickedness. Just then, up out of Somerset came an old man with a rheumy eye, a gammy leg and a long white beard.

'Old man,' called out the Devil, 'how far to Devizes?'

The old boy saw in a wink who he was talking to. 'Well, sir,' he replied, 'funny you should ask that, because I'm a stranger here, and I'm going to Devizes myself. When I set off out of Somerset I was a little boy of ten, and I ain't never got to Devizes yet.'

'*That* far?' roared the Devil in a taking, and with that he stamped away back to Hell with his sack, leaving the pile of earth dumped out by the roadside. And that's Cley Hill.

A pile of remarkable stories attaches to this hill of modest height and aspect. Crop circles, UFOs, beams of light shining out of the earth. Ancient Palm Sunday rituals involving beating the bounds, burning the grass to scare the Devil out of the hill, knocking a ball from bottom to top with a curved stick in a mysterious game called Bandy. Men of Somerset and men of Wiltshire meeting here once a year, to strive together with staves and fists for the right to style themselves masters of the county border. And a holy healing spring, the Hog's Well, in which weakling piglets could be cured, but whose waters must never be drunk by humans or cattle for fear of being swallowed up in the ground.

Cley is a shape-shifter. Coming to it the way we normally do, from the east, it sits like a cardinal's biretta, with a crowning burial mound as the bobble on top. But from the southern side, as Jane and I walk north along the lane with the hill lifting before us, it suddenly sprouts a stepped flank and abrupt quarry cliffs. Now we can get a proper idea of the Iron Age ramparts that encircle the summit, and the impressive steepness and bulk of the hill. It's a rise of only two hundred and fifty feet from foot to crown, but the damp chalk grassland slope makes a slippery climb, especially if you're shod in boots with worn-down treads as I invariably seem to be. I've climbed Cley Hill in snow, in fog, in sunshine and drizzle, and I've slipped and skidded every single time. This Boxing Day is no exception. I'm over on my backside within seconds. The melodeon in the rucksack on my back doesn't help my balance, of course. But the Cley Hill ritual demands its presence.

Today's company includes my swift-striding sister Lou and two of my daughters. The third is engaged in her own Boxing Day rituals elsewhere. We're getting blown around. It's a tremendously windy morning, the sky entirely clouded, the clouds driving north-east at forty or fifty miles an hour. But it's not cold, not at all. The thermometer's at fifteen degrees, ridiculously high for a Boxing Day. There has been no crispness to the days this month, no frost at night. Daffodils are out at Windsor Castle, we hear; apple trees in blossom, primroses in my mother's Somerset garden. An unsettling feeling of spring is in the air, before the old year has even turned.

The cyclically recurring band of warm water in the Pacific Ocean whose climatic effects go by the name of El Niño is making its presence felt from eight thousand miles away. It's set to be the most disruptive El Niño event on record. The surface of the Pacific is 3°C warmer than normal. Warm air pushing north has caused a huge bend in the jet stream – the band of 200 mph winds blowing west to east some seven miles overhead in the upper atmosphere – making it swoop south of the British Isles and carry its freight of Atlantic depressions towards and then across these islands. Worldwide, temperatures have been soaring. In the UK, it has been the warmest December ever, at 8°C average, 4°C above the norm. There'll be no snow on Cley Hill this year. It's been the wettest December ever, too, the rainstorms impressing themselves on the mind not so much for their duration as their ferocious intensity. The Met Office has begun to give each storm a name, American-style, working through the alphabet.

Abigail, Barney, Clodagh, Desmond, Eva – the names have conferred personality on each storm, making the details of each one easier to remember, but they also carry a kind of jokey menace, as though these destructive forces were hitherto cheerful neighbours who'd suddenly begun threatening to kick your front door down. The TV forecast map has shown them approaching day after day, each successive tempest rushing in from the Atlantic in a solid blue curve with green and yellow patches indicating pulses of particular concentration, to empty an unprecedented load of water over the hills and uplands of southern Scotland and Cumbria.

These catastrophic monsoon-like rains in the north have been dominating the news all month. On 4–5 December more than thirteen inches of rain fell on the Honister Pass in the Lake District, more than would normally be expected in the whole month. The village of Glenridding has been flooded four times in quick succession; then Carlisle, York, Hebden Bridge, Leeds, as the saturated ground sheds its excess water into rivers that can't cope. The rivers overtop flood defences that are unfitted for this scale of emergency. Everyone becomes familiar with the intimate images of personal disaster – an armchair floating down a village street, a shopkeeper drearily sweeping floodwater out of her premises, a woman weeping for tiredness and misery as the river invades her living room for the third time in a week. Politicians wade about in green wellingtons and promise money, locals rail at the Environment Agency and the Government. Meanwhile, the reality of man-made global warming is still a topic of fierce debate in the media and on the internet. People don't want to believe it, and many of them don't and won't.

We're not the only people on the hill today. Cley Hill is a popular post-Christmas pudding-shaker, and walkers in ones and twos are beginning to set out. The wind buffets us all from behind, driving us upwards like a ruthless sergeant. Near the top of the hill the green button of the summit barrow comes into view. I climb to the dome-shaped roof of the barrow and brace myself against the gusts, wiping away wind tears to admire the view. The long line of Mendip to the north. The flanking downs of Salisbury Plain in the east. Flatter fields westwards into Somerset. To the south, the rise of thickly wooded hills into Longleat's parkland. And nearer at hand, the remarkable shape of Cley Hill revealed, a scoop in its north-east face smoothly sculpted out of the chalk by the wind and rain of ten millennia. The north face of the hill falls very steeply to a narrow neck and the opposing rise of the out-lying Little Cley Hill where my two daughters are already cavorting far below.

Old stories tell of a great standing stone that had its place on top of the barrow. It must have been toppled by the time of another tale that spoke of a likeness of the Devil's face on the underside – but woe betide anybody foolhardy enough to turn the stone over and take a look. Later the stone was tumbled down the hillside by a group of men. The barrow is also known to be the home of Bogley, guardian spirit of the Hog's Well. He can be called forth at dusk in the shape of a dwarf by anyone bold enough to walk nine times widdershins around the barrow.

The depression in the crown of the bowl barrow has deepened since last Boxing Day. It's tempting to imagine it as the socket of the demonic standing stone, but in fact it was Sir Richard Colt Hoare of nearby Stourhead House who left this hollow behind after he excavated the 4,000-year-old barrow early in the nineteenth century, finding 'some ears of wheat undecayed, and fragments of pottery, charred wood, and ashes'. No dwarf, though. There's another bowl barrow beyond, and an Ordnance Survey trig pillar, but the summit of Bogley's barrow is where I play my melodeon on Boxing Day. Today that proves impossible. I literally cannot stand up. As soon as I unship my rucksack the wind rocks me and knocks me sideways, shoving me over the edge and down the side of the barrow. I make an undignified slide, and fetch up in a heap in the ditch below. The melodeon in its hard case proves undamaged. I seat myself in comparative shelter on the grassy flank of the barrow, get the instrument out of its green velvet nest and put it across my knee.

All my life I have longed to get a tune, a proper tune, out of something. As a child I found I couldn't read the dots. Violin and piano were tortured for a few sessions and let go. Playing the guitar is all very well, but I've been doing that for fifty years and I'm still only a plucker and strummer. Playing the mouth-organ doesn't really count. What I wanted was something simple and effective, for playing simple and effective stuff – morris tunes, Irish traditional tunes, English folk songs, Christmas carols. An instrument you could play in a session without spoiling the tune. The melodeon, much beloved of morris musicians for its portability and jolly loudness, seemed the business. It is a fairly unsophisticated type of squeeze-box. It's small and handy, and you play it by fingering a row of buttons, not a piano-style keyboard with its school-room associations of bafflement and failure. Watching a barrel-bodied, horse-laughing, ale-quaffing melodeon mangler spill out tunes for the Mount Bures Morris Men outside a Suffolk pub, the ignoble thought first occurred to me: if a bloke like that can play one of those, then I bloody well can.

My musical friend Dave Richardson gave me a crude old Hohner to practise on. I played it upside down for the first month I had it. Then Dave gently took it off me and reversed it. Oh, I see. The right hand plays the tune, eh? Not like a guitar, then. So what does the left hand do? 'I'd cut the left hand out completely,' advised my other musical friend, Andy, 'till you've got some sort of control over the right.' Ah.

The melodeon across my knee on Cley Hill this Boxing Day is a tiny but good one, a Castagnari Lilly. It was a pearl wedding present from Jane, on account of its pearl buttons and fretwork hearts. I love it, and her – probably not in that order. My youngest daughter has had great fun on social media this Christmas, being pictured clapping her hands over her ears and pulling a death-ray face whenever I get out the Castagnari. But she has stridden ahead with her sister, and is far below on the low summit of Little Cley Hill. I can just make out her pink coat against the rain-leached green of the turf. I have my own sister Lou beside me, and Jane in sight by the trig pillar. I flick off the two straps that hold the bellows of the melodeon together. A good puff of air into the lungs of the instrument, an experimental twiddle of the buttons, and it's the 'Floral Dance' that wobbles out across the hill and is blown to shreds by the wind.

Sometimes we dance on Cley Hill, sometimes we sing. It depends who is up here with us, the mood we're in and the state of the weather. Today seems quieter, in the dead of the year. 'Speed The Plough' fades out in a squabble of misplaced notes. Then my cold fingers find the opening notes of 'Early One Morning'. That gentle, lilting tune cannot help but bring my father to mind. It was one of a clutch of old English folk songs that he would occasionally sing to amuse himself and us – 'Blow The Wind Southerly', 'Sweet Polly Oliver', 'My Boy Billy'. The rich and lugubrious voice of Kathleen Ferrier was his gold standard, and her throbbing contralto planted those simple songs in a mulch of melancholy in my childish imagination. They've stayed rooted there ever since. I often find myself humming them and picturing Dad. 'Early One Morning' was one of his favourites. So I sit and watch the Mendip Hills, his native ground, and play it for myself, for the day, to please dwarf Bogley, and for the shade of Dad that sits alongside my sister and me.

Dad died at home, in the dead of the year 2005. In September he was diagnosed with terminal cancer, and in November he died. It was that quick.

For a year he'd been shedding weight, losing his appetite and his capacity for conversation. I went with him to learn his fate from the oncological specialist, but Dad couldn't hear what was being said. I had to break it to him when we got home. 'Am I dying?' was his forthright question. I ducked that; I couldn't bring myself to give him a straight answer. But he knew anyway.

He needed help to eat, to wash at the tiny dressing-room basin, to put his clothes on. He staggered when he stood up, reeled when he walked. We all chipped in to help him. It was heart-wrenching sometimes, doing for this suddenly aged and decrepit man what he had done so willingly for me at the other end of life. Of course Shakespeare's Seven Ages of Man came often to mind. There was no mewling and puking as Dad completed his circle of life; he was just too proud for that. But the shrunken shank, the big manly voice turning to piping and whistling, the 'sans teeth, sans eyes, sans taste' were all present and correct. Yet sometimes there was a ghost of humour, too. There was a moment just before the end when I was helping him to wash and finding it especially difficult to bear, this witnessing by his own son of the childlike helplessness of such a private man. He caught my eye, and although by then he was incapable of laughing there was a lift of his brows and a twitch of the bow-shaped ends of his mouth that signalled an internal quake of amusement at his – and my – predicament.

We did manage one last walk together in that final month of his life. He wanted a breath of the winter air, but was struggling to summon the will to walk. At last we got out into the street, Dad in muffler, cap, gloves and the disreputable Canadian quilted jacket he wore for gardening. Every step was an effort of will. His boots dragged along the road, and he leaned on my arm with his old ash stick in his left hand for balance. We shuffled to the corner and round it; but that was as much as Dad could do. A last look up the lane towards the familiar buildings of the village, and then we turned about laboriously and inched back to the house. I felt like Turner's little black steam tug in *The Fighting Temeraire*, guiding the ghost of the once-great ship of the line to its dissolution.

We buried Dad in the village churchyard next to his parents and his grandparents. I happened to be the last to walk away up the church path, and as I did so there was a rush and swoosh through the air over my head. A crowd of starlings swooped over the churchyard wall and alighted in a tree. For a minute or so they perched there fluffing out their feathers, chattering and whistling in the bare branches, overflowing the heart of the tree, the only lively and joyful things anywhere in sight. Then with a tremendous screeching they all took off together and went streaming away to the west. It felt like a release, and I carried it away with me.

Up on Cley Hill the wind pushes and blusters. I watch my daughters climb back up the northern slope, striding upwards with energy to rejoin us, ready to tease me. But I've pre-empted them, and the melodeon is safely and silently back in its case. We go our separate ways down Cley Hill, and meet up again at its southern foot for the splashy quarter-mile back to the car. The second half of the Boxing Day ritual has yet to be commenced, and we can't be late for the Mummers.

The old pub hasn't changed a bit. It never does. From one year's end to the next, the only day I step inside its dark blue front door is Boxing Day. I scarcely rank as a regular. But this plain and unassuming public house vies with the Hunter's Lodge near Priddy as my favourite pub in the whole wide world. That's the strength of the magic generated here every 26 December.

The simple, square-built alehouse has stayed true to itself, in a manner that's almost incredible when set against the way that other such rural pubs have been done up as fancy eateries, or have been closed and converted into handsome private dwellings. A purchaser could make a beautiful house out of it, if they had the chance. But they haven't had, up till now. The former landlords ran the place exactly as they wanted for more than fifty years, and when their son took over a couple of years ago he made his intention very plain – to keep the pub as a proper rural Somerset ale and cider house.

There are two public rooms, one on either side of the short stone-flagged passage that leads to the glass-windowed pulpit of the bar counter. Furniture is extremely solid and unadorned – hefty, iron-bound wooden tables, wooden settle benches round the walls that are hard on less generously padded bottoms. The venerable wallpaper has been tanned to a soft, ashy brown by decades of tobacco smoke. There's a coal fire in winter. No piped music; live music only. No extensive blackboard menu; in fact, no food at all, except for the occasions where everyone brings something and there's a sort of communal picnic. Cider and beer from the barrel at prices that would make a Londoner blink. Conversation earthy enough to make him blink some more. And an ambience that is very hard to put a finger on, but one you'd never find in a more sophisticated kind of place. Put it like this: if this pub were an old acquaintance, it would be the one that you rely on to tell you the unvarnished truth, the yardstick friend you measure all others by.

It's hard to say which is the parlour and which the public bar. The square room on the left of the passage is where we're headed. We get our drinks and go in to claim a place. First come, first seated is the rule. I glance round, and it's all right. Everything is just as it was, last Boxing Day and all the Boxing Days before that, including the faded Christmas decorations that must have glittered like anything when they were first put up back in the 1950s. The melodeon case is beside me on the bench, and I lift out the Castagnari and put it across my knee for a couple of experimental twiddles. But it's not really time for music yet. People are chatting in family groups. Talk takes precedence, so I sip my beer and joke with the others until John Salmon and his melodeon show up.

*

If there's a guardian spirit of Boxing Day at the old pub, it's John. Morris dancer, countryman and exceptionally dry merry-andrew, he acts as MC in these frolics, finding a place with modest good humour for anyone to sing, no matter how donkey-like their braying. He, too, plays a Castagnari Lilly, and what particularly endears him to me is that, when it comes to playing the melodeon, the two of us have contrived over the past decade to stay neck and neck, to remain exactly as good, or bad, as each other. For me that makes John an ideal musical companion, especially today, as I suspect he might have nosed a little ahead over the past few months and I'm therefore very slightly on my mettle. We sit side by side, clacking the buttons and aspirating the bellows of our diminutive machines, squeezing out a rustic stew of morris tunes. 'General Monk's March'. 'The British Grenadiers'. 'Princess Royal', a favourite of mine with its strange rhythms and elongated bits. 'Let's have something we can sing!' someone calls, and we respond. Nothing properly learned or practised, just what issues from the tips of our fingers. 'Clementine'. 'Leaving Of Liverpool'. 'Drink Up Thy Zider' – Adge Cutler of the Wurzels, who wrote this hymn to Somerset's champagne, is a local folk god. Some of the later verses are pure guesswork, but no one here knows them any better than we do, so nobody objects. A red-faced cider drinker with a kingly moustache sits nearby, and he trolls out the choruses with as much come-all-ye gusto as a planted laugher in a TV comedy-show audience. People join in, not uproariously, but with a sort of wry enjoyment.

John and I squeeze out the last drops of 'Drink Up Thy Zider', and he looks across and tips me the wink. Here it comes. We pack the melodeons away and weave between the standing drinkers – every seat is taken by now, and more people are coming in, some standing in the fireplace, others looking in at the door under one another's arms. Four or five furtive-looking characters follow us outside, and we slink round to the pub's backyard like a bunch of ne'er-do-wells, getting funny looks from a party as they go in the door.

In the yard stands a mud-splashed car, its boot lid wide open to reveal the filthiest heap of rags imaginable. A fusty, musty smell hangs in the drizzly air. No one in their right mind would plunge their hands, as we do, deep in among these disgusting scraps. Each man drags out a tattered suit of ribbons and struggles into it. Mine smells particularly rank, testament to the sweat of past Boxing Days. The ribbons tickle my arms and hang down over my face. I lean into the boot and extract a circular shield with a faded motif painted on it, a crescent and star. A tiny red felt fez completes my ensemble. I am transformed, if not transfigured, and am become Bullslasher, the Turkish Knight, with a morris dancer's stick for a scimitar. All round me similar characters are garbing and arming themselves into existence – Father Christmas in fur-rimmed hat, St George with a great red cross on his shield, Little Johnny Jack the sweep with a family of jigging, sinister little dolls sewn at the back of his rag tunic, and John magnificently metamorphosed into Doctor Dodd, half medic and half magician, with a black bag full of bloody teeth and dubious potions. Little Johnny Jack seizes a drum, and to its death-march beat we pace solemnly round the yard and in at the blue front door.

The Mummers are here.

Wikipedia knows the score. 'Mummers Plays are seasonal British folk plays, performed by troupes of amateur actors known as mummers or guisers (or by local names such as *rhymers, pace-eggers, soulers, tipteerers, wrenboys, galoshins, guysers,* and so on.' Tipteerers and galoshins: I'm proud to be numbered among them. How long have the wrenboys and guysers of Britain been at these merry, menacing capers? The mummers' plays have little to do with the sacred miracle plays performed from medieval times onwards, though they are at least as old in origin. They feature ritualistic battles between good and evil – St George versus the Dragon, or, as in the case of the Boxing Day play that's enacted at the old pub, St George versus the Turkish Knight. There's a resurrection theme involved. But there's a pungent whiff of the pagan, too, emanating from the mummers and their ceremonies. How else to view Little Johnny Jack and the clutch of shrunken, blank-faced homunculi he's burdened with, or Doctor Dodd, half demonic presence and half resurrection man?

We enter the bar room at the slow march, the drum hollowly beating out the step. The reaction of the crowd on first sight of us is divided according to age. The grown-ups all break out smiling and laughing, and some raise a cheer. The children stare in shock, and some begin to whimper, at the tall shapes smothered in ribbons, the faces blanked out by rags and tatters, the hands that grasp the big sticks. We shuffle to our places between fireplace and tables, and stand awkwardly facing the people. Father Christmas takes a deep breath that shivers his synthetic beard. Are you sitting comfortably? Then we'll begin.

> *'In come I, Old Father Christmas, Welcome or*
> *Welcome not;*
> *I hope Old Father Christmas will never be forgot.*
> *Christmas comes but once a year, and when it comes,*
> *it brings good cheer,*
> *Roast beef, Plum pudding and Good English beer.'*

I can hardly see a thing through the fusty rags dangling over my face. The fez feels as though it's about to slip off. At least I won't be stumped for the few words I have to say – they are pasted on the back of my round shield. I stare at them as Father Christmas gives tongue again.

> *'Oh! Room for a gallant soldier. Oh! Room, room,*
> *give him rise,*
> *And I'll show you the best activity as seen this*
> *Christmas tide!'*

St George inhales hard and takes a tighter grip of his stick as Father Christmas intones:

> *'Now, if you don't believe what I do say,*
> *Step in, St George, and show the way.'*

St George's number is up. He takes a stride forward and brandishes his shield with its painted cross.

> '*In come I, St George, that noble champion bold;*
> *With my glittering sword I won five crowns of gold.*
> *'Twas I that fought the fiery Dragon and brought him*
> *to the slaughter,*
> *And by fair means won Sheba, the King of Egypt's*
> *daughter.'*

That is quite a boast, St George. The Dragon we all know about. But winning Sheba, whether by fair means or by foul? Where did that notion spring from? The Queen of Sheba visited King Solomon with her 'very great train, with camels that bare spices, and very much gold, and precious stones' more than a thousand years before your time on earth. Maybe it was some other Sheba. Whatever the truth of that, I'm going to have to respond to your bloodthirsty challenge:

> '*Bring me that one, who will before me stand,*
> *And I will cut him down, with sword in hand.'*

A last furtive peer through the rag blindfold at the words on the back of my shield, and I spring forward and roar:

> '*In come I, the Turkish knight; Bullslasher is my*
> *name;*
> *With sword and buckler by my side I mean to win the*
> *game!*

I mean to win the game, St George, and lay you in the
    mud (bloodcurdling laugh) –
So first I draw my scimitar, and then thy precious
    blood.'

I flourish my morris stick and turn my back on St
George, playing the stage Turk for all I'm worth as I twirl
an imaginary three-foot-long moustache. From a murky
corner of the child inside the man, my father's voice
murmurs in my ear: 'Pipe down, old boy . . . take a back
seat . . . don't show off.' But I fancy there's the ghost of
Dad's laugh in there, too. There is no knowing when this
particular mummers' play came into being, but it's certainly rife with causes for offence in modern times, if any
offence-takers should happen to be present in this room.
Dad had enough to do with the reality of the Turks and
Turkey in his day, bolstering the British presence in the
secret world there, visiting signal intelligence stations
near the Russian border, dealing with the fallout from the
'Ünye incident' in 1972 when three men on GCHQ duties
were abducted from their living quarters in northern
Turkey and killed by the Turkish People's Liberation
Army. What would he have made of the farrago of naked
stereotyping and religious chauvinism that informs the
Nunney Mummers' Play? How would he react to his son's
present histrionics? Today, at the end of this year of keeping company with the man I've grown to love, I hope and
think he'd have laughed like a drain.

St George and I exchange more cock-fighting insults,
and then we fall to blows. *Whack! Smack!* The morris

sticks batter on the hardboard shields. My ridiculous fez slips over one eye. 'St George and the Turkish Knight fight,' instructs the script, ' – St George falls.' He's got to do that without hurting himself, but equally it needs to be a proper clattering tumble. Down he goes, like a good'un, and measures his length among the crisp packets. I stamp and guffaw round my fallen foe. Everyone boos. I shake my scimitar stick at the room. More boos. A dog growls, and a four-year-old scrambles onto his mother's knee with a quivering lower lip. Hmm, should I tone it down a bit? At this moment during last Boxing Day's revels I reversed my stick and sprayed St George with imaginary bullets. Today I cannot possibly do anything of the sort. Even as I brandish the round shield with its crescent and star triumphantly over the body of England's patron saint, a thought has slipped like a thief into my brain: is anyone thinking of what happened six weeks ago, out in the wider world? The piles of young dead in a Parisian concert hall; the Islamist fanatics, just as young, who killed those strangers, and then killed themselves. Is all this play-acting in the worst possible taste just now? But in the next instant, through the curtain of tatters over my face, I see the laughing faces and the gleeful enjoyment in this roomful of tipteerers and galoshins, and I recall what our Boxing Day ritual is all about – a vigorous shout in the face of winter, a rude snook cocked at gloom and death.

Father Christmas steps forward and makes lament in undisciplined metre:

'Oh! Bullslasher, what hast thou done? – killed Old
    England's Son!
Through the head and through the knee –
Ten thousand pounds will not cure he . . .
Is there a Doctor to be found?'

Yes, there is a Doctor to be found. Up steps the cadaverous and unsettling figure of Doctor Dodd in a voodoo-esque bowler hat. His demand of fifty pounds for resurrecting the dead man is swiftly beaten down by Father Christmas to five farthings. Where has the disquieting Doctor come from? Italy, Spitaly, Germany and Spain, all around the world and back again. 'What can'st thou cure?' Father Christmas enquires.

'The Itch, the Stitch, the Palsy and the Gout,
Pains within, and Pains without,
If the Devil's in, I'll fetch him out.'

And we can believe that he will, as he opens his necromantic black bag. Out come the monstrous pincers and the giant bloody tooth (shrieks from the children). Out comes the bottle marked 'Poison'.

'What have I got in this bottle?
I'll pour some down his thrittle-throttle.'

Down goes the medicine with a fearsome gargling. Doctor Dodd scrabbles round in his bag once more like a fairground huckster, looking for 'pills to cure all ills – the

Itch, the Stitch, the money grubs and the burlybubs'. Out comes a pill that remarkably resembles a ping-pong ball, which he inserts between the lips of the fallen hero (who promptly blows it ceiling-high). Then the devilish Doctor declaims:

'Rise up, St George, and fight once more.'

And miraculously, St George does. I have been watching and grinning along like everyone else, and am almost caught out. I jerk myself forward and trade a few more blows with my revitalized opponent. Perhaps in earlier and more full-blooded times the Mummers' Play would have ended with the Turkish Knight done to death, St George and the cross of Christ raised in triumph to universal acclaim and yelling. But not this play, not on this day. Father Christmas steps between us, calling a halt to hostilities – though I notice that he does leave our options open with the way he puts his point: 'Stop this fight without delay, and fight it out another day.' Round goes the hat, and out we troop again to the muffled beat of the drum, out to the yard to recover our everyday selves in the drizzly half-light as we doff our funny headgear and pull our rank costumes over our heads.

The frolic ends as the frolic must, with songs and music and another glass of beer. My daughters prove not to have been terminally embarrassed by their old man making a spectacle of himself behind his mask of rags. No face, no name, no pack drill: the convention seems to have survived another year.

'"The Holly And The Ivy",' says John to me. 'Unaccompanied.' We settle our melodeons back into their velvet nests and click the cases shut. We stand side by side among all those in the room, singing the familiar words to an old tune I never hear sung except in this one pub on this one day of the year. The carol is a perfect blend of the sacred and the pagan, hymning the greenwood holly with its berries as red as blood, its bark as bitter as gall, its prickles as sharp as a crown of thorns. There is no sentimental optimism here. We get a clear-eyed appraisal of what is, and what is to be. There is no reprieve in store for sweet Jesus Christ and his mother, no easy way out for us; only the foreshadowing of pain and a promise of salvation, the cycle of the months, the hope of release, and the ongoing story of the berries, the prickles, the bark and the blossom.

Along the Way for Ever

And the January man comes round again
in woollen coat and boots of leather,
To take another turn and walk along
the icy road he knows so well;
The January man is here for starting
each and every year
Along the way for ever.

New Year's Eve at Lyme Regis. The circle of the year is closing with a battle between sea and land.

Lyme Bay is roaring. I have never seen anything like it. The sea has come alive, its pulse translated into arched ridges that run from side to side of the bay as they march in on the town. Down in the shallows a guillemot trails a broken wing and silently awaits its fate. Spray thickens the air, flying over the forked arm of the Cobb breakwater in great smoky sheets. The sky over the land is a deep, calm blue. Giant white clouds, creased with silver, drift ponderously in across the coast where the cliff face of Golden Cap has been gilded with fresh falls.

The Cobb has been set in the sea here for eight hundred years, but it looks as though the sea must crush it today. People stand on the beach and marvel. The sea punches at the snaking granite wall, shaking the beach with deep vibrations. Each wave comes in with smooth deliberation, seeming to gather speed before it disappears behind the old customs shed at the fork of the breakwater. There's a two-second pause; then a geyser of white water shoots up behind the shed, forty feet into the air, spreading into a monstrous fan or sea-god profile before the wind catches the spray and whips it to rags. But this is only a shake of the fist from the wave, whose main body surges on sideways to engulf the western half of the Cobb. A collar of yellow foam unfurls along the top of the wall, running landward as a crumpled mass before sluicing back down the outer edge in a line of cascades. A foolhardy photographer is huddled at the turn of the breakwater, silhouetted against each wave burst. He must have climbed the irregular flight of steps called Granny's Teeth and wedged himself at the top. He'll be lucky to escape with only a drenching.

The isle is full of noises. Herring gulls complain as they are tossed about the sky. In the boatyard the wind conducts a discordant symphony. Halyards scream and rap out a voodoo tattoo against the aluminium yacht masts, whose hollow tubes emit a tremulous hooting as though owls were being woken inside. The canvas boat-covers drum madly. It all fades under the tremendous noise of the sea, rhythmically grinding its pebbles as I walk away along the strand of Monmouth Beach, where the waves have reduced the leathery ribbons of oarweed to pimpled stubs like octopus suckers. Under the charcoal-black slope of Ware Cleeves I find a seat on a skeleton tree that has fallen from the cliff edge. You can't sit long with impunity under Ware Cleeves; they are the most unstable cliffs in Britain, and last night's stormy high tide and seeping rain have already pulled out fresh fans of dark gleaming clay. But this is the place from which to watch the Cobb embattled in foam, shaking off the spray as it defies the sea that is outflanking it, slowly and implacably, on its ageless mission to recapture the land.

A year ago I sat with my son George on this tree, or another like it, and watched my three-year-old grandson defy the sea. George lives in tropical north-west Australia on a coast of vast white beaches and turquoise shallows. But he still carries a torch from time to time for the cold grey skies and seas of his native land. Lyme Regis had been his favourite place when he was small, a beach and sea to thrill a little boy. He longed for his own son, Arthur, deep in the grand obsession with dinosaurs and fossils that grips small boys, to love the idiosyncrasies of Lyme as he himself once had.

This New Year's Eve I sit alone, gazing up the empty beach. Arthur stands in my mind's eye as I saw him last December, grasping two fistfuls of 'fossils' – stones and shells – in hands red with wind. George stalks beyond in familiar mode, tall and thin, head bent as he searches the cliff foot for ammonites. Arthur heaves with giggles. He runs forward to the water's edge and hurls one handful of stones into the sea, then the other, shouting 'Go back, sea!' at the top of his voice. Then he skips backwards as the wave splashes his shoes. He trips and falls on his back. The intrepid warrior bursts into tears. But I don't need to run to his aid. His father is there to pick him up, brush him down and hold him in his arms and kiss him better. Arthur blinks and swallows. He bends for more stones and runs to confront the sea again. 'Go back, sea!'

It is partly painful, partly joyful, this film of my absent son and grandson that plays in the skull cinema today. And there is someone else to share it with, of course, another presence who comes quietly to sit on the skeleton tree and watch over the two figures with me.

This time of year, ten years ago almost to the day, I

saw my father in the icy road outside his house, a month after he had died. It was only a glimpse as I turned the corner, a momentary impression of the old man, a little stooped, in his fawn corduroys and shabby old quilted jacket – Dad, and not Dad. He was setting off up the village street, hobbling as he had done for the last few months of his life. I had not even time for a pang of the heart, or a shout, before the impression had faded out to nothing. But it left me gasping and bereft.

Now under Ware Cleeves I feel him at my shoulder. Through the snowy hills of Shropshire and the spring flowers of Teesdale, the midsummer dusk of Shetland and the Wiltshire mists and mires of autumn on the Harroway I have cursed him and argued with him, turned my back and closed my ears to him. I have missed him and learned to love him better. Now on the salt-crusted tree by the sea I fold him in my arms, like a father or a son, eager to hear what happens next in the endlessly unfolding story of the rising sun and the running deer.

Acknowledgements

I have great pleasure in thanking ...

My parents John and Elizabeth Somerville for letting me run freely around the flood country of the River Severn as a small boy.

My sisters Julia and Lou for reading through *The January Man* and making helpful suggestions.

Michael Herman, Peter Marychurch, Robin Gilbert, John Silcock and Michael Canning for help and information on my father's time at GCHQ.

Sgt Wheatley of Coombe Hill police station, for very sensibly giving me a ticking-off instead of sending me to Borstal.

Nick Godden, warden at RSPB Marshside Reserve in Lancashire, for taking the time to walk me round and share his knowledge.

The hospitable and helpful residents of Foula, including Fran Dyson Sutton and her partner Magnus Gear, Kenny and Bobby Gear, Sheila and Jim Gear, Penny, Paul, Robert and Jack Gear, Davie Henry and Bryan Taylor.

George and David Hoyles, and Julian and Stafford Proctor, farmers of Long Sutton, Lincs, for taking time out of their busy harvest schedules to show me over their farms on the reclaimed shores of the Wash.

Roger Wilkins of Land's End Farm, Mudgley, and John Salmon of Smithwick's Bridge, Marston Bigot, for helping to keep proper Somerset traditions alive and kicking.

Susanna Wadeson and the Transworld team for their inspiring enthusiasm.

Carry Akroyd (carryakroyd.co.uk) for so brilliantly encapsulating the twelve months of 'The January Man' in her beautiful jacket artwork. And many thanks, too, to Richard Shailer for his evocative chapterhead sketches.

Ben Mason and Michael Alcock for helping it all past the post.

Penny Grigg for her tireless keyboard feats.

My dear brother-in-mischief, Andrew Finlay.

My wife Jane for being my constant and observant companion.

And very special thanks to Dave Goulder (davegoulder.co.uk) for generously allowing me to quote from his wonderful song 'The January Man'. Neither he nor I has been able to determine the copyright-holder, but I will be glad to hear from them.

Bibliography

February

A Night in the Snow by the Revd E. Donald Carr, reissued Dodo Press (www.dodopress.co.uk) – available from Burway Books, Church Stretton (www.burwaybooks.co.uk)

The Golden Arrow by Mary Webb, reissued Echo Library, 2008

April

Walking in the Lake District by H. H. Symonds (W. & R. Chambers, 1933)

A Pictorial Guide to the Lakeland Fells by A. Wainwright (Frances Lincoln, 2005–9)

May

Journey Through Britain by John Hillaby (Constable, 1968)

The Everlasting Mercy by John Masefield (1911), reproduced by permission of the Society of Authors as the Literary Representative of the Estate of John Masefield

A Pennine Way Companion by A. Wainwright (Frances Lincoln, 2012)

July

Ballad of Thomas the Rhymer: *Minstrelsy of the Scottish Border, Part Three* by Sir Walter Scott (Edinburgh University Press, 2016). Online: *https://ebooks.adelaide.edu.au/s/scott/walter/ minstrelsy-of-the-scottish-border/chapter68.html*

The Complete Works of Bede, trans. by J. A. Giles (Delphi Classics, 2015)

Music: *On Cheviot Hills* by Alistair Anderson & The Lindsays (alistairanderson.com)

August

Kenzie the Wild Goose Man by Colin Willock (André Deutsch, 1972; Boydell Press, 1985)

The Eye of the Wind by Peter Scott (Hodder & Stoughton, 1961)

September

The Complete Poems of Samuel Taylor Coleridge (Penguin Classics, 1997)

The Fairy Caravan by Beatrix Potter (reissued Warne, 2011)

The Feathered Man by Jeremy De Quidt (Random House Children's Publishers, 2012)

The Toymaker by Jeremy De Quidt (David Fickling Books, 2010)

'Robyn Hode and the Munke': http://www.sacred-texts.com/neu/eng/child/ch119.htm

October

The Most Amazing Places in Britain's Countryside copyright © The Reader's Digest Association, Inc. Reprinted with permission.

November

Ancient Trackways of Wessex by H. W. Timperley and Edith Brill (Nonsuch Publishing, 2005)

Round About a Great Estate by Richard Jefferies (Smith, Elder & Co., 1880)

A Shepherd's Life by W. H. Hudson (Cambridge University Press, 2011)

Layamon's *Brut*
Old English: http://quod.lib.umich.edu/cgi/t/text/text-idx?c=
cme;idno=LayCal
Modern English: http://www.gutenberg.org/cache/epub/
14305/pg14305-images.html

If you have been
inspired by these walks,
you can follow Christopher's
footsteps by downloading a free
walking guide from his website

christophersomerville.co.uk/walkingguide

Christopher Somerville is the walking correspondent of *The Times*. He is one of Britain's most respected and prolific travel writers, with thirty-six books, hundreds of newspaper articles and many TV and radio appearances to his name. He lives in Bristol.